T0192868

The ionized material that constitutes plasma permeates almost all of the universe. This book describes the linear theory of many different waves and instabilities that may propagate in collisionless space plasmas. Electrostatic and electromagnetic fluctuations, and a variety of instablility sources, are considered.

Applications of the theory are discussed with respect to spacecraft observations in the solar wind, terrestrial magnetosheath, magnetosphere and magnetotail and at the bow shock and magnetopause.

Tables at the ends of most chapters summarize wave and instability nomenclature and properties, and problems for the reader to solve are interspersed throughout the text. Together these make this book of great value to both the student and the research worker in space physics.

Cambridge atmospheric and space science series

Theory of space plasma microinstabilities

Cambridge atmospheric and space science series

Editors

Alexander J. Dessler
John T. Houghton
Michael J. Rycroft

Titles in print in this series

M. H. Rees, *Physics and chemistry of the upper atmosphere*
Roger Daley, *Atmospheric data analysis*
Ya. L. Al'pert, *Space plasma, Volumes 1 and 2*
J. R. Garratt, *The atmospheric boundary layer*
J. K. Hargreaves, *The solar-terrestrial environment*
Sergei Sazhin, *Whistler-mode waves in a hot plasma*
S. Peter Gary, *Theory of space plasma microinstabilities*

Theory of space plasma microinstabilities

S. Peter Gary

Los Alamos National Laboratory

CAMBRIDGE UNIVERSITY PRESS
Cambridge, New York, Melbourne, Madrid, Cape Town, Singapore, São Paulo

Cambridge University Press
The Edinburgh Building, Cambridge CB2 2RU, UK

Published in the United States of America by Cambridge University Press, New York

www.cambridge.org
Information on this title: www.cambridge.org/9780521431675

First published 1993
This digitally printed first paperback version 2005

A catalogue record for this publication is available from the British Library

Library of Congress Cataloguing in Publication data

Gary, S. Peter.
Theory of space plasma microinstabilities / S. Peter Gary.
p. cm. — (Cambridge atmospheric and space science series)
Includes index.
ISBN 0–521–43167–0
1. Space plasmas. 2. Plasma instabilities. I. Title.
II. Series.
QC809.P5G27 1993
523.01–dc20 93–3649 CIP

ISBN-13 978-0-521-43167-5 hardback
ISBN-10 0-521-43167-0 hardback

ISBN-13 978-0-521-43748-6 paperback
ISBN-10 0-521-43748-2 paperback

Contents

Preface

If a charged particle species of a collisionless plasma possesses a non-Maxwellian velocity distribution function, a short wavelength normal mode of the system may grow in amplitude. This is a microinstability; its theory is well described by the Vlasov equation. The purpose of this monograph is to describe in an accurate way the theory of damped normal modes and a limited number of microinstabilities that may arise in various space plasma environments.

The two words that best characterize the work described in this book are "limited" and "accurate." In order to keep the discussion limited, I have chosen idealized, not observed, distribution functions. Many spacecraft have provided excellent observations of electron and ion distributions in the Earth's magnetosphere and nearby solar wind. The tremendous variety of these distributions makes it difficult to select a few for special representation. My choice here has been to use Maxwellian or bi-Maxwellian distributions with field-aligned drifts to represent some of the more important general free energy sources. Although the resulting instabilities may not correspond to any particular data set, I hope that each one represents the general properties of a very broad class of data.

To provide accuracy, I have followed the same procedure for each distribution function and plasma model. After assuming a zeroth-order distribution, I derive (or at least explicitly state) the associated dispersion equation without approximation. Because I deal with linear theory throughout this book, it is always straightforward to do this, although the algebra gets tiresome at times. A complete Vlasov dispersion equation involves the sums (and often products) of transcendental functions of a complex variable. Such an equation can be a rich source of physics, but there are often subtleties and surprises along the path to its solution. To make sure that the proper physics is attained throughout this book I numerically solve the complete linear dispersion equation for the waves and instabilities of that dispersion equation and then, when appropriate, use the exact results to justify reductions to analytic forms that simplify the physical interpretation.

This procedure stands in contrast to some of the literature in this field that invokes analytic approximations before computers are brought to bear on the problem. The danger in this procedure is that, because one is dealing with a transcendental equation, inappropriate approximations can change its character, introducing spurious roots or changing the character of true solutions. Because my preference is not to make use of references that may be subject to such errors, you may find some of your favorite citations missing from my list of references. Although I do cite references concerning computer simulations and spacecraft observations, the primary emphasis throughout this book is on linear theory.

This book began as a series of lecture notes that I used to supplement textbook material for an advanced plasma physics course I taught at the University of New Mexico, Los Alamos, in 1981. I gradually added material over the years until 1987, when I became Leader of the Space Plasma Physics Group at the Laboratory. The press of administrative responsibilities then caused the manuscript to be neglected for several years. But the recent departure of my younger child to college has opened up more evening and weekend time for me to work on the book once again; a slight nudge from the good folks at Cambridge University Press and the realization that I had better get the job done before I lost what few plasma insights remained have spurred me to complete this work.

I thank all my colleagues in plasma theory and experiment who, through their unselfish scientific collaborations with me, have substantially contributed to the insights and interpretations found in this book. Among these many individuals I would like to single out Dan Winske, who has been particularly generous, cooperative, and persistent in supporting my efforts to understand the theory of space plasma microinstabilities. And I especially thank my wife, Pat, for her patience and understanding in allowing me to put in the time after hours that has enabled me to, finally, finish this manuscript.

The writing of this book was performed under the auspices of the US Department of Energy. The research described here has been supported by funding from both the DOE Office of Basic Energy Sciences, Division of Engineering and Geosciences, and several programs of the National Aeronautics and Space Administration, including the NASA Space Plasma Theory Program.

S. Peter Gary
Los Alamos, NM

1
Introduction

Plasma instabilities are normal modes of a system that grow in space or time. Thus the word "instability" implies a well-defined relationship between wavevector **k** and frequency ω; this in turn implies that the associated plasma fluctuations are relatively weak so that linear theory is appropriate to describe the physics.

This book uses linear Vlasov theory to describe the propagation, damping and growth of plasma modes. Linear theory cannot describe the ultimate fate of a plasma instability, nor its interactions with other modes. Of course the questions of how an instability reaches maximum amplitude, whether and how it contributes to plasma transport and whether such transport affects the overall flow of mass, momentum and energy at large scales are crucial for establishing the relevance of microphysics to large scale modelling of space plasmas. But these questions must be addressed by nonlinear theory and computer simulation, which are beyond the purview of this book. Our relatively modest goal is to use computer solutions of the unapproximated Vlasov dispersion equation to firmly establish the properties of plasma normal modes; our hope is that this information will provide a useful foundation for the interpretation of computer simulations and spacecraft observations under conditions of relatively weak fluctuation amplitudes.

1.1 Micro- vs macro-

The most general classification of growing modes in a plasma divides them into two broad categories: macroinstabilities at relatively long wavelengths and microinstabilities at shorter wavelengths. Macroinstabilities depend on the configuration-space properties of the plasma and are well described by fluid equations; microinstabilities are driven by the departure from thermodynamic equilibrium of the plasma velocity distributions and therefore must be described by the Vlasov equation or other kinetic equation. In a magnetized plasma with the gyroradius of a characteristic ion a_i, macroinstabilities generally grow most rapidly at $ka_i \ll 1$, whereas microinstabilities generally have maximum growth rates at $ka_i \gtrsim 1$.

The distinction between the two categories is not clear cut; macromodes may have appreciable growths at short wavelength and micromodes may persist to small wavenumbers. The distinction is further blurred by the fact that the wavenumber at maximum growth rate of some instabilities depends on the plasma parameters and may slide from the micro- to the macro- regime as the parameters change. Nevertheless, the macro- vs micro- distinction is a convenient one to begin any general discussion of unstable plasma modes.

1.2 The kinetic equation

Most plasma microphenomena are thought to be well described by Maxwell's equations coupled with a kinetic equation; this equation determines the time development of $f_j(\mathbf{x}, \mathbf{v}, t)$, the jth species distribution function. On the other hand, plasma macrophenomena often are described adequately by the field equations coupled with a set of fluid equations that may, under suitable assumptions, be derived from a kinetic equation. Fluid equation variables are functions only of space and time ($\mathbf{x}$ and t).

The velocity-space information of a kinetic equation implies that the physics of microphenomena is much more diverse than that associated with macrophenomena. The most important new plasma property associated with kinetic plasma physics is wave-particle interactions. This is the capability of waves with appropriate phase speeds or group velocities to exchange energy with plasma particles moving with the same velocities. Wave-particle interactions are exemplified by Landau and cyclotron damping (discussed in Chapter 2) and by plasma instabilities driven by non-Maxwellian properties of the velocity distribution (Chapters 3, 4, 7 and 8).

For the nonrelativistic plasmas considered in this book, the general form of the kinetic equation is (Montgomery and Tidman, 1964; Nicholson, 1983)

$$\frac{\partial f_j}{\partial t} + \mathbf{v} \cdot \frac{\partial f_j}{\partial \mathbf{x}} + \frac{e_j}{m_j}\left(\mathbf{E} + \frac{\mathbf{v} \times \mathbf{B}}{c}\right) \cdot \frac{\partial f_j}{\partial \mathbf{v}} = \left(\frac{\partial f_j}{\partial t}\right)_{collision} \qquad (1.2.1)$$

where e_j is the charge of the jth species particle, m_j is the mass of the jth species particle, and the right-hand side represents the effect of some unspecified collision term. We use Gaussian cgs units throughout this book because they are the most widely used units in the research literature.

A concise but thorough discussion of the relationship between cgs and rationalized mks units is given as an Appendix in Boyd and Sanderson [1969].

In many of the models used in this book, we assume that the plasma consists of two species, electrons (denoted by subscript e) and ions (subscript i); the latter will usually be taken to be protons (subscript p). At times we will consider one of these species to consist of two distinct components; we will assume that the distribution function of each component satisfies Equation (1.2.1). In such cases the subscript notation will be introduced as appropriate. However, the e and i subscripts will denote overall electron and ion properties throughout the book. The notation $\sum_j$ will denote a summation over all components and species in the plasma.

The distribution function is related to the macroscopic plasma quantities through velocity moment integrals. The particle density of the jth species or component is obtained from the zeroth velocity moment:

$$n_j \equiv \int d^3v\, f_j. \tag{1.2.2}$$

From the first velocity moment we define the particle flux density of the jth species or component

$$\mathbf{\Gamma}_j \equiv \int d^3v\, \mathbf{v} f_j, \tag{1.2.3}$$

the momentum density $\mathbf{P}_j \equiv m_j \mathbf{\Gamma}_j$, and the drift velocity $\mathbf{v}_{dj} \equiv \mathbf{\Gamma}_j / n_j$. If there is a background magnetic field in the $\hat{\mathbf{z}}$-direction, the field-aligned component of the drift velocity will be denoted by $\mathbf{v}_{oj} \equiv \hat{\mathbf{z}} \Gamma_{zj} / n_j$, and the cross-field drift by $\mathbf{v}_{\perp dj}$.

From the second moments of the velocity we define the kinetic energy density tensor of the jth species or component

$$\mathbf{W}_j \equiv \frac{m_j}{2} \int d^3v\, \mathbf{v}\mathbf{v} f_j$$

with

$$W_j \equiv \frac{m_j}{2} \int d^3v\, v^2 f_j$$

and the temperature of the jth species or component

$$T_j \equiv \frac{m_j}{3n_j} \int d^3v\, (\mathbf{v} - \mathbf{v}_{dj})^2 f_j$$

with separate temperatures parallel and perpendicular to the background magnetic field

$$T_{\parallel j} \equiv \frac{m_j}{n_j} \int d^3v\, (v_z - v_{oj})^2 f_j$$

and

$$T_{\perp j} \equiv \frac{m_j}{2n_j} \int d^3v \, (\mathbf{v}_\perp - \mathbf{v}_{\perp dj})^2 f_j.$$

The third velocity moment yields the kinetic energy flux density, or, as it is more commonly known, the heat flux density of the jth species or component:

$$\mathbf{q}_j \equiv \frac{m_j}{2} \int d^3v \, \mathbf{v} v^2 f_j.$$

Maxwell's equations for the electric and magnetic fields are

$$\nabla \cdot \mathbf{E} = 4\pi\rho \qquad (1.2.4)$$

$$\nabla \times \mathbf{E} = -\frac{1}{c}\frac{\partial \mathbf{B}}{\partial t} \qquad (1.2.5)$$

$$\nabla \times \mathbf{B} = \frac{1}{c}\frac{\partial \mathbf{E}}{\partial t} + \frac{4\pi}{c}\mathbf{J} \qquad (1.2.6)$$

$$\nabla \cdot \mathbf{B} = 0 \qquad (1.2.7)$$

where the charge density $\rho = \Sigma_j e_j n_j$ and the current density $\mathbf{J} = \Sigma_j e_j \mathbf{\Gamma}_j$ are defined in terms of f_j by Equations (1.2.2) and (1.2.3). Equations (1.2.1) and (1.2.4) through (1.2.7) are the basic equations for the kinetic theory of microinstabilities.

Problem 1.2.1. Prove that, if the collision term conserves the particle number of each species, velocity integration of the kinetic equation (1.2.1) yields the equation of continuity

$$\frac{\partial n_j}{\partial t} + \nabla \cdot \mathbf{\Gamma}_j = 0. \qquad (1.2.8)$$

1.3 The Vlasov equation

Detailed derivations of the kinetic equation and various forms of the collision term of Equation (1.2.1) in a fully ionized plasma are given in Montgomery and Tidman (1964), Clemmow and Dougherty (1969) and Nicholson (1983), for example. In this book we are concerned with plasmas that are sufficiently hot and/or tenuous so that the number of particles within a sphere of Debye length radius is large:

$$n\lambda_D^3 = n_e \left(\frac{T_e}{4\pi n_e e^2}\right)^{3/2} \gg 1. \qquad (1.3.1)$$

This is called a collisionless plasma. In this case, the right hand side of

Equation (1.2.1) is negligible and one has the Vlasov equation

$$\frac{\partial f_j}{\partial t} + \mathbf{v} \cdot \frac{\partial f_j}{\partial \mathbf{x}} + \frac{e_j}{m_j}\left(\mathbf{E} + \frac{\mathbf{v} \times \mathbf{B}}{c}\right) \cdot \frac{\partial f_j}{\partial \mathbf{v}} = 0. \qquad (1.3.2)$$

This is the kinetic equation used throughout this book.

Problem 1.3.1. By integrating appropriate velocity moments of the Vlasov equation, derive the following momentum and energy equations:

$$\frac{\partial \boldsymbol{\Gamma}_j}{\partial t} + \nabla \cdot \mathbf{W}_j = \frac{e_j}{m_j}\left(n_j \mathbf{E} + \frac{\boldsymbol{\Gamma}_j \times \mathbf{B}}{c}\right), \qquad (1.3.3)$$

$$\frac{\partial W_j}{\partial t} + \nabla \cdot \mathbf{q}_j = \frac{e_j}{m_j} \boldsymbol{\Gamma}_j \cdot \mathbf{E}. \qquad (1.3.4)$$

Why does the magnetic field not appear in the energy equation?

1.4 Definitions

Whenever it is present, the background magnetic field is assumed to be constant, uniform, and pointing in the z-direction: $\mathbf{B}_o = \hat{\mathbf{z}} B_o$, with $B_o > 0$. The subscripts $\parallel$ and $\perp$ refer to directions parallel and perpendicular to $\mathbf{B}_o$, respectively; thus, for example, $\mathbf{v}_\perp = \hat{\mathbf{x}} v_x + \hat{\mathbf{y}} v_y$. The unperturbed plasma is assumed to be charge neutral with a total electron density n_e. Throughout this book the following notation will be used:

Thermal speed of the jth component: $v_j \equiv \left(\frac{T_{\parallel j}}{m_j}\right)^{1/2}$

Plasma frequency of the jth component: $\omega_j \equiv \left(\frac{4\pi n_j e_j^2}{m_j}\right)^{1/2}$

Cyclotron frequency of the jth species: $\Omega_j \equiv \frac{e_j B_o}{m_j c}$

Debye wavenumber of the jth component: $k_j \equiv \left(\frac{4\pi n_j e_j^2}{T_{\parallel j}}\right)^{1/2}$

(Signed) gyroradius of the jth component: $a_j \equiv \frac{v_j}{\Omega_j}\left(\frac{T_{\perp j}}{T_{\parallel j}}\right)^{1/2}$

Beta of the jth component: $\beta_j \equiv \frac{8\pi n_e T_{\parallel j}}{B_o^2}$

Inertial length of the jth component: $\frac{c}{\omega_j}$

Plasma beta: $\beta \equiv \frac{8\pi \Sigma_j n_j T_{\parallel j}}{B_o^2}$

Alfvén speed: $v_A \equiv \left(\frac{B_o^2}{4\pi n_e m_p}\right)^{1/2}$

The Boltzmann factor k_B is always understood to multiply the temperatures T_j. Note that β_j and v_A are defined in terms of the total electron density, n_e, not the jth component density n_j.

Problem 1.4.1. Assume representative solar wind parameters $n_e \simeq 10\ \mathrm{cm}^{-3}$; $T_e \simeq T_p \simeq 10^5$ K; and $B_o \simeq 10^{-4}$ G. Calculate the electron and proton thermal speeds and the Alfvén speed, and compare them with a representative solar wind speed of 400 km/sec. Similarly, calculate and compare representative electron and ion plasma and cyclotron frequencies in the solar wind.

A reduced distribution function, which is useful for one-dimensional pictures, is defined as an integral over the velocity components perpendicular to some direction:

$$f_j(v_z) \equiv \int dv_x dv_y f_j(\mathbf{v}).$$

Charge neutrality, which is assumed in zeroth order for all situations, requires

$$\rho = \sum_j e_j n_j = 0. \tag{1.4.1}$$

In addition, for all configurations except those explicitly associated with a current, we will assume that an electric field acts to maintain zero current and in zeroth order

$$\mathbf{J} = \sum_j e_j n_j \mathbf{v}_{dj} = 0. \tag{1.4.2}$$

Throughout this book we consider weak plasma fluctuations. This means that the fluctuating fields are sufficiently small in amplitude that both the fields and the distribution functions may be expanded

$$\begin{aligned} f_j(\mathbf{x}, \mathbf{v}, t) &= f_j^{(0)}(\mathbf{x}, \mathbf{v}) + f_j^{(1)}(\mathbf{x}, \mathbf{v}, t) + f_j^{(2)}(\mathbf{x}, \mathbf{v}, t) + ... \\ \mathbf{E}(\mathbf{x}, t) &= \mathbf{E}_o(\mathbf{x}) + \mathbf{E}^{(1)}(\mathbf{x}, t) + \mathbf{E}^{(2)}(\mathbf{x}, t) + ... \\ \mathbf{B}(\mathbf{x}, t) &= \mathbf{B}_o(\mathbf{x}) + \mathbf{B}^{(1)}(\mathbf{x}, t) + \mathbf{B}^{(2)}(\mathbf{x}, t) + ... \end{aligned} \tag{1.4.3}$$

where superscript (j) represents a quantity proportional to $|\mathbf{E}^{(j)}|$.

1.5 Fluctuations, waves and instabilities

The traditional development of the linear theory of instabilities in collisionless plasmas follows a well-established procedure: The linear Vlasov equation is subjected to a Fourier/Laplace analysis in space/time, yielding fluctuating particle densities and particle flux densities that are inserted into Maxwell's equations (1.2.4) through (1.2.7) to yield a dispersion equation. The solution of this dispersion equation relates frequency ω and wavevector $\mathbf{k}$ and thereby determines the normal modes of the plasma.

The dispersion equation may be solved either as a boundary value problem (ω is given as real, and one solves for a complex component of $\mathbf{k}$) or as an initial value problem ($\mathbf{k}$ is given as real, and one solves for a complex ω). The latter approach is subject to fewer mathematical ambiguities and is the approach more often followed in the literature; we follow it exclusively.

Thus throughout this book the complex frequency will be $\omega = \omega_r + i\gamma$ where γ is the growth or damping rate. We regard as a heavily damped oscillation any solution of the linear dispersion equation that satisfies $\gamma < -|\omega_r|/2\pi$. We use the term *waves* to describe those weakly damped solutions that satisfy $-|\omega_r|/2\pi \leq \gamma \leq 0$, and describe as *instabilities* growing solutions with $\gamma > 0$. And *fluctuations* will denote both stable waves and instabilities. The *phase speed* of a fluctuation, the speed at which a point of constant phase of a single mode propagates through the plasma, is ω_r/k. Although the observed frequency of a plasma fluctuation is a function of the relative motion between the observer and the medium bearing the wave, the damping or growth rates γ calculated from homogeneous plasma theory are independent of the frame in which the calculation is performed.

Problem 1.5.1. Show that an observer moving with velocity $\mathbf{v}_o$ with respect to a plane wave of frequency ω_r and wavevector $\mathbf{k}$ observes a frequency $\omega_r - \mathbf{k} \cdot \mathbf{v}_o$. This change in the observed frequency is called the *Doppler shift*.

Unless stated otherwise, the wavevector will be taken to lie in the (y, z)-plane, so that

$$\mathbf{k} = \hat{\mathbf{y}}k_y + \hat{\mathbf{z}}k_z. \tag{1.5.1}$$

Subscripted wavenumbers represent components and in analytic expressions may assume either positive or negative values. The wavenumber magnitude $k = (k_y^2 + k_z^2)^{1/2}$ will be understood to be always positive. Our numerical evaluations will consistently use $k_y \geq 0$ and $k_z \geq 0$; we will reverse wave propagation direction by reversing the sign of ω_r. The angle betweeen $\mathbf{k}$ and $\mathbf{B}_o$ will be denoted by θ; thus $\hat{\mathbf{k}} \cdot \hat{\mathbf{B}}_o = \cos\theta$. The maximum growth rate over the full range of wavevectors associated with an instability will be denoted by γ_m; the wavevector corresponding to maximum growth will be $\mathbf{k}_m$.

If the distribution functions of each plasma species are Maxwellian and no external electric fields are present, the dispersion equation typically yields nongrowing roots. In order to yield one or more plasma instabilities, the dispersion equation must be based on distribution functions involving *free energy*; that is, some non-Maxwellian property corresponding, for example, to an anisotropy or an inhomogeneity.

As the free energy (say a relative drift speed between two components) is increased, the imaginary part of the frequency, γ, of a damped mode becomes less negative until $\gamma = 0$ is reached at some wavevector. We term this condition the *threshold* of the associated instability because a further increase of free energy leads, at some wavevectors, to $\gamma > 0$, that is, wave growth. At and somewhat above threshold, it is often true that at least one component (j) is resonant with the instability; i.e. $|\zeta_j| \lesssim 1$ where ζ_j is the argument of some plasma dispersion function $Z(\zeta_j)$ used in the linear dispersion equation. In this regime, wave growth depends on velocity-space details of the jth component distribution function and the instability is termed *kinetic*.

If, as the free energy is further increased (e.g. the relative drift speeds of the components become much greater than the component thermal speeds), the maximum growth rate also continues to increase, all plasma components often become nonresonant ($|\zeta_j| \gg 1$), and the dispersion equation can be reduced to a cold plasma form. In this regime, the growing mode is usually termed a *fluid instability*.

Given a particular source of free energy, a plasma may be unstable to several different modes. So the classification of any microinstability requires identification of both the free energy and the dispersion properties. Although at times it will be necessary to defer to historical precedent, we will, as much as possible, identify microinstabilities described in this book by both their free energy and dispersion. Thus, for example, the kinetic instability with ion acoustic dispersion and driven by the electron/ion relative drift speed will be called the electron/ion acoustic instability. As the electron/ion relative drift speed increases and the unstable mode becomes fluid-like, the instability is more appropriately called the electron/ion two-stream instability.

When discussing plasma waves and instabilities, it is convenient to separate their fluctuating electric fields into two types: longitudinal ($\mathbf{k} \times \mathbf{E}^{(1)} = 0$) and transverse ($\mathbf{k} \cdot \mathbf{E}^{(1)} = 0$). The complete solution of the general dispersion equation will typically have contributions from both types of fields; we define these as $\mathbf{E}_L^{(1)} = \mathbf{kk} \cdot \mathbf{E}^{(1)}/k^2$ and $\mathbf{E}_T^{(1)} = \mathbf{k} \times \mathbf{E}^{(1)}/k$, respectively.

Plasma fluctuations that have only a longitudinal electric field may be derived through the use of a kinetic equation such as (1.2.1) and a single Maxwell equation: Poisson's equation (1.2.4). Such waves and instabilities

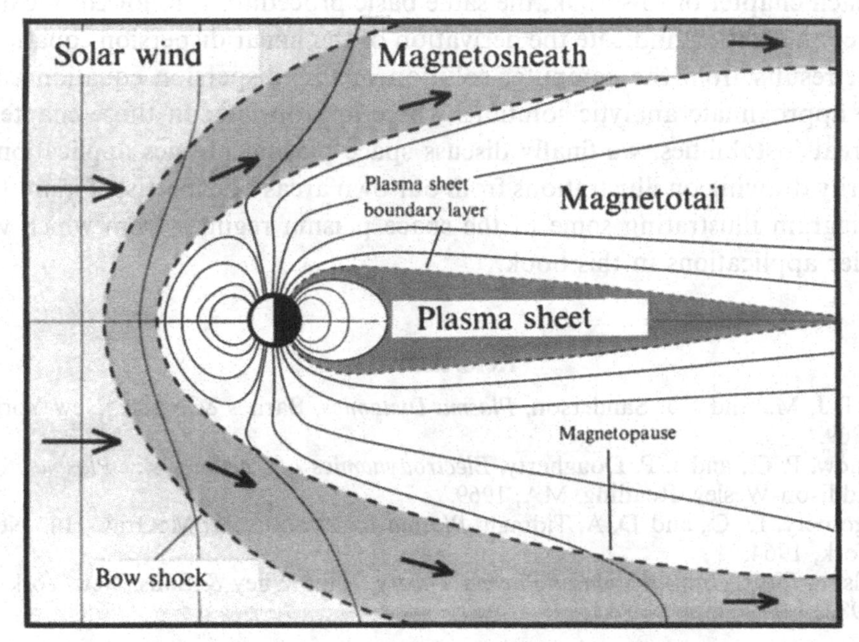

Fig. 1.1 Some space plasma regimes near the Earth.

have $\mathbf{B}^{(1)} = 0$ and are usually called "electrostatic." In contrast, waves and instabilities with fluctuating electric and magnetic fields perpendicular to the wavevector and with no longitudinal electric field can be described through the use of an appropriate kinetic equation, Faraday's equation (1.2.5) and the Ampère-Maxwell equation (1.2.6). These fluctuations are sometimes called "electromagnetic."

Most fluctuations in space plasmas of nonzero β have both transverse and longitudinal components. To provide a consistent terminology throughout this book, we will use the following nomenclature: fluctuations with only a longitudinal component will be called "electrostatic;" fluctuations with only transverse electric fields will be termed "strictly electromagnetic." If the fluctuating field energy of a mode satisfies

$$0 < |\mathbf{E}_T^{(1)}|^2 + |\mathbf{B}^{(1)}|^2 < |\mathbf{E}_L^{(1)}|^2$$

we will term the wave "primarily electrostatic," and if

$$0 < |\mathbf{E}_L^{(1)}|^2 < |\mathbf{E}_T^{(1)}|^2 + |\mathbf{B}^{(1)}|^2$$

we will term the mode "primarily electromagnetic." Finally, the term "electromagnetic" will encompass fluctuations with arbitrary ratios of longitudinal and transverse fluctuating fields.

In each chapter of this book, the same basic procedure is followed. We define a configuration, indicate the derivation of the linear dispersion equation, exhibit results from the computer solution of this dispersion equation and derive approximate analytic solutions where appropriate. In those chapters that treat instabilities, we finally discuss space plasma physics applications, primarily drawing on illustrations from our own areas of expertise. Figure 1.1 is a diagram illustrating some of the space plasma regimes from which we consider applications in this book.

References

Boyd, T. J. M., and J. J. Sanderson, *Plasma Dynamics*, Barnes & Noble, New York, 1969.

Clemmow, P. C., and J. P. Dougherty, *Electrodynamics of Particles and Plasmas*, Addison-Wesley, Reading, MA, 1969.

Montgomery, D. C., and D. A. Tidman, *Plasma Kinetic Theory*, McGraw-Hill, New York, 1964.

Nicholson, D. R., *Introduction to Plasma Theory*, John Wiley & Sons, New York, 1983.

2
Electrostatic waves in uniform plasmas

In this chapter we examine electrostatic fluctuations in homogeneous, collisionless plasmas. We further assume that the distribution functions describing the individual plasma components are isotropic. In the absence of any inhomogeneity or anisotropy in the distributions, the imaginary part of the wave frequency satisfies $\gamma \leq 0$ and there are only "waves," that is, normal modes of the plasma that do not grow in time or space.

We consider collisionless plasmas in which the evolution of the distribution function of the jth component is described by the Vlasov equation (1.3.2). We also consider only low β plasmas in which the electrostatic approximation is valid so that the fluctuating fields are described by Poisson's equation (1.2.4). In Section 2.1 we define the Maxwellian used as the zeroth-order distribution function throughout this chapter. Section 2.2 considers electrostatic waves in unmagnetized plasmas, Section 2.3 examines electrostatic waves in magnetized plasmas, and Section 2.4 is a brief summary. Throughout this chapter, we assume that the background plasma is charge neutral, has zero current, and bears no steady-state electric field.

2.1 Zeroth-order distributions: the Maxwellian

We here assume there are no anisotropies or inhomogeneities in the zeroth-order distribution, so we choose $f_j^{(0)}$ to be independent of position and the direction of $\mathbf{v}$. For numerical examples in this chapter, we will use Maxwellian distributions:

$$f_j^{(0)}(\mathbf{v}) = f_j^{(M)}(v) = \frac{n_j}{(2\pi v_j^2)^{3/2}} \exp\left(-\frac{v^2}{2v_j^2}\right). \tag{2.1.1}$$

We use Maxwellian distributions not only here in Chapter 2, but also in our study of electromagnetic waves in Chapter 6. Furthermore, we use distributions related to the Maxwellian, e.g. the drifting Maxwellian, in chapters which examine instabilities, such as Chapter 3.

Although distribution functions observed in space plasmas are not typically observed to be strictly Maxwellian, they are frequently Maxwellian-like; that is, they are relatively smoothly varying with a single maximum and decrease very rapidly with increasing speeds above some characteristic thermal speed. We use Maxwellian distributions because they have the additional advantage that they yield finite velocity moment integrals for all powers of the velocity; in particular, they yield well-defined results for all important physical quantities including the density, particle flux density, temperature, and heat flux density defined in Section 1.2.

2.2 Electrostatic waves in unmagnetized plasmas

The linear Vlasov equation in a magnetic-field-free plasma is

$$\frac{\partial f_j^{(1)}(\mathbf{x},\mathbf{v},t)}{\partial t} + \mathbf{v}\cdot\frac{\partial f_j^{(1)}(\mathbf{x},\mathbf{v},t)}{\partial \mathbf{x}} + \frac{e_j}{m_j}\mathbf{E}^{(1)}(\mathbf{x},t)\cdot\frac{\partial f_j^{(0)}(\mathbf{v})}{\partial \mathbf{v}} = 0. \tag{2.2.1}$$

In the electrostatic approximation, the fluctuating electric field is given by Poisson's equation (1.2.4).

Landau (1946) solved these equations as an initial-value problem in a homogeneous medium, carrying out a Fourier transform in $\mathbf{x}$ and a Laplace transform in t. He showed that, at asymptotically large times, the fluctuating potential is proportional to $\exp(-i\omega t)$, where the complex Laplace transform variable ω is related to the real Fourier variable $\mathbf{k}$ through a dispersion equation $\omega = \omega(\mathbf{k})$. Landau also showed that, for a Maxwellian electron distribution function, the imaginary part of ω is negative, so that the fluctuating potential decreases in time. This is what has become known as *Landau damping*.

The mathematically rigorous formulation of the Laplace transform initial-value problem has been discussed at length in many plasma physics textbooks, particularly Montgomery and Tidman (1964) and Montgomery (1971). In this chapter and, indeed, in this book we are concerned only with the asymptotic solutions that emerge from the dispersion equation. To obtain this dispersion equation, it is sufficient to assume that all first-order quantities vary in space and time as

$$h^{(1)}(\mathbf{x},t) = h^{(1)}(\mathbf{k},\omega)\exp[i(\mathbf{k}\cdot\mathbf{x} - \omega t)] \tag{2.2.2}$$

where the components of $\mathbf{k}$ are taken to be real but the frequency is complex: $\omega = \omega_r + i\gamma$.

Under the electrostatic approximation, the fluctuating electric field may be written as the gradient of the electric potential:

$$\mathbf{E}(\mathbf{x}, t) = -\nabla\phi(\mathbf{x}, t) \tag{2.2.3}$$

so that Equation (2.2.1) becomes

$$-i(\omega - \mathbf{k}\cdot\mathbf{v})f_j^{(1)}(\mathbf{k}, \mathbf{v}, \omega) = \frac{e_j}{m_j}\phi^{(1)}(\mathbf{k}, \omega)\ i\mathbf{k}\cdot\frac{\partial f_j^{(0)}(\mathbf{v})}{\partial \mathbf{v}}.$$

Following Equation (1.2.2), we obtain the first-order density of the jth component by integrating the first-order distribution function of the jth component over velocity:

$$n_j^{(1)}(\mathbf{k}, \omega) = \frac{e_j}{m_j}\phi^{(1)}(\mathbf{k}, \omega)\int\frac{d^3v}{\mathbf{k}\cdot\mathbf{v} - \omega}\mathbf{k}\cdot\frac{\partial f_j^{(0)}(\mathbf{v})}{\partial \mathbf{v}}. \tag{2.2.4}$$

It is convenient to define a function that characterizes the response of the jth component density to the fluctuating potential. Thus we define the jth component susceptibility $K_j(\mathbf{k}, \omega)$ as

$$n_j^{(1)}(\mathbf{k}, \omega) = -\frac{k^2\phi^{(1)}(\mathbf{k}, \omega)}{4\pi e_j}K_j(\mathbf{k}, \omega) \tag{2.2.5}$$

so that

$$K_j(\mathbf{k}, \omega) = -\frac{\omega_j^2}{n_j k^2}\int\frac{d^3v}{\mathbf{k}\cdot\mathbf{v} - \omega}\mathbf{k}\cdot\frac{\partial f_j^{(0)}(\mathbf{v})}{\partial \mathbf{v}} \tag{2.2.6}$$

or, integrating by parts in the component of velocity parallel to $\mathbf{k}$,

$$K_j(\mathbf{k}, \omega) = -\frac{\omega_j^2}{n_j}\frac{\partial}{\partial\omega}\int\frac{d^3v f_j^{(0)}(\mathbf{v})}{\mathbf{k}\cdot\mathbf{v} - \omega}. \tag{2.2.7}$$

Poisson's equation is linear, so use of Equation (2.2.5) yields the linear dispersion equation

$$1 + \sum_j K_j(\mathbf{k}, \omega) = 0. \tag{2.2.8}$$

The linear theory of plasma waves and instabilities is often reduced to such an equation that determines the dispersion properties, that is, the relationship between ω and $\mathbf{k}$ for the plasma modes. Using Equation (2.2.6) with Equation (2.2.8), the linear dispersion equation can be written

$$1 - \sum_j\frac{\omega_j^2}{n_j k^2}\int\frac{dv_k}{v_k - \omega/k_k}\frac{\partial f_j^{(0)}(v_k)}{\partial v_k} = 0 \tag{2.2.9}$$

where k_k bears a subscript as a reminder that it may be either positive or negative, in contrast to k which is understood to be always nonnegative.

Before Equation (2.2.9) can be solved, a crucial issue must be addressed. The integrand has a singularity at $\mathbf{k} \cdot \mathbf{v} = \omega$; thus, the contour of integration for the component of velocity parallel to $\mathbf{k}$ (denoted by v_k) must be specified. Under the assumption that $k_k > 0$, the correct prescription was given by Landau (1946): this contour must pass below the singularity, whatever the value of γ.

The choice of the Landau contour for the v_k integration coupled with the assumption of real $\mathbf{k}$ permits Equation (2.2.8) to yield solutions for $\omega(\mathbf{k})$ that are in general complex. From Equation (2.2.2), $\gamma < 0$ corresponds to a wave that decays in time, while $\gamma > 0$ corresponds to a growing mode, an instability. This Landau damping or Landau growth associated with $\gamma \neq 0$ is the single most important consequence of the Vlasov equation.

If the mode in question is unstable, the singularity of Equation (2.2.9) lies in the upper half v_k-plane, and an integration along the real v_k-axis satisfies the Landau prescription. However, if Equation (2.2.9) is to yield a stable or damped wave with $\gamma \leq 0$, the Landau contour must involve an additional contribution from integration around the singularity; the result is a function of ω that is the analytic continuation of the function in the unstable case. Thus, if the reduced distribution $f_j^{(0)}(v_k)$ is analytic for all $|v_k| < \infty$, then $K_j(\mathbf{k}, \omega)$ is an analytic function of ω for all finite values of the complex frequency and has all the advantageous properties that term implies (Montgomery and Tidman, 1964, Section 5.3).

If $|\gamma| \ll |\omega_r|$ (which is not always true), one may approximate the Landau contour by a straight line along the v_k-axis that excludes the point $v_k = \omega_r/k_k$ and a semicircular path around the pole at $v_k = \omega/k_k$ (Montgomery and Tidman, 1964):

$$\int \frac{dv_k}{v_k - \omega/k_k} g(v_k) = \mathrm{P} \int \frac{dv_k}{v_k - \omega_r/k_k} g(v_k) + i\pi g(v_k)|_{v_k=\omega_r/k_k} \tag{2.2.10}$$

where $g(v_k)$ is an arbitrary function of v_k, and $k_k > 0$.

Further analytic reductions depend on the value of ω_r/kv_j, the ratio of the phase speed of the wave to the thermal speed of the jth component. We will address specific solutions as we consider individual modes; here we simply note that, for sufficiently small γ, the real part of Equation (2.2.9) yields an estimate for ω_r and the imaginary part of the same equation implies

$$\gamma = -k_k \pi \frac{\sum_j \frac{\omega_j^2}{n_j} \frac{\partial f_j^{(0)}(v_k)}{\partial v_k}\Big|_{v_k=\omega_r/k_k}}{\sum_j \frac{\omega_j^2}{n_j} \mathrm{P} \int \frac{dv_k}{(v_k-\omega_r/k_k)^2} \frac{\partial f_j^{(0)}(v_k)}{\partial v_k}} \tag{2.2.11}$$

(where $k_k > 0$.). Thus the crucial elements in determining the sign of γ are the slopes of the reduced distribution functions at $v_k = \omega_r/k_k$.

Equation (2.2.11) and its generalizations to magnetized plasmas and electromagnetic waves are often useful in providing qualitative estimates of damping and growth rates. However, the quantitative dependence of the growth rate on plasma parameters is not obvious from this equation; moreover, its limitations due to the approximations necessary to derive it are not clear.

It is important to recognize the complex transcendental nature of the linear dispersion equation and the many pitfalls associated with obtaining correct and general analytic solutions from that equation. Thus, the point of view taken throughout this book is that the safest and most general approach to the dispersion equation is to solve it on a computer, and let the numerical results guide the subsequent analytic approximations.

If we assume that the zeroth-order distribution is Maxwellian (Equation (2.1.1)) then the susceptibility becomes

$$K_j(\mathbf{k}, \omega) = -\frac{k_j^2}{2k^2} Z'(\zeta_j) \tag{2.2.12}$$

where $Z(\zeta)$ is the plasma dispersion function and $\zeta_j = \omega/\sqrt{2}|k_k|v_j = \omega/\sqrt{2}kv_j$. The plasma dispersion function appears frequently in dispersion equations derived from distributions that are Maxwellian or Maxwellian-like. The properties of this important function are briefly described in Appendix A.

Problem 2.2.1. Show that Equations (A.1), (A.2) and (A.3) are equivalent forms of the plasma dispersion function. Then use these forms to prove the following properties of $Z(\zeta)$:

$$Z'(\zeta) = -2[1 + \zeta Z(\zeta)] \tag{A.4}$$

$$Z(\zeta^*) = -[Z(-\zeta)]^* \tag{A.5}$$

$$Z(\zeta) + Z(-\zeta) = 2i\sqrt{\pi}\exp(-\zeta^2) \tag{A.6}$$

Problem 2.2.2. The use of absolute value on k_k in the definition of ζ_j is a

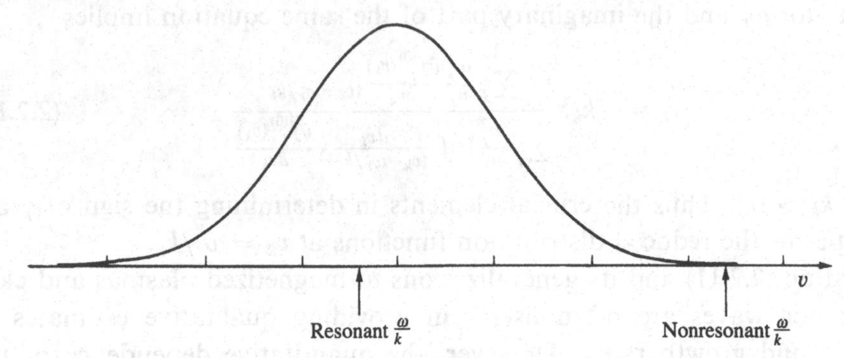

Fig. 2.1 A Maxwellian reduced velocity distribution with a resonant phase speed and a nonresonant phase speed.

nontrivial result. Prove that Equation (2.2.12) is valid for both positive and negative values of k_k.

If $|\zeta_j| \lesssim 1, f_j^{(0)}$ has an appreciable value at the $\omega_r = \mathbf{k} \cdot \mathbf{v}$ resonance of Equation (2.2.7), and the jth component is said to be resonant with the wave. Particles moving with velocities that satisfy this condition are said to be in Landau resonance with the wave; for such particles $\mathbf{E}^{(1)} \cdot \mathbf{v}$ remains a constant (if the perturbing effects of $\mathbf{E}^{(1)}$ on $\mathbf{v}$ are ignored) and there can be a strong exchange of energy between the wave and those particles. If $|\zeta_j| \gg 1$, the Landau resonance lies well beyond the range of thermal velocities of the jth component, there are relatively few particles that experience the Landau resonance and that component is said to be nonresonant.

These ideas are illustrated in Figure 2.1, which shows a Maxwellian velocity distribution function and two phase speeds. The "resonant ω/k" arrow corresponds to $\zeta_j \simeq -1$; there are many particles of this component that are moving near the phase speed of the wave, and there are strong wave-particle interactions here. The "nonresonant ω/k" arrow corresponds to $\zeta_j \gg 1$; the wave is moving faster than almost all the particles of the distribution and the wave-particle interaction is correspondingly much weaker here.

2.2.1 Electron plasma waves

At frequencies above the electron plasma frequency, the ions are unable to respond to the fluctuating fields and it is only the lighter species that contributes to electron plasma waves (Landau, 1946). Figure 2.2 shows a representative dispersion plot of this wave. At $k \ll k_e, \zeta_e \gg 1$ and the electrons are nonresonant. In this limit, the asymptotic expansion of $Z'(\zeta_e)$

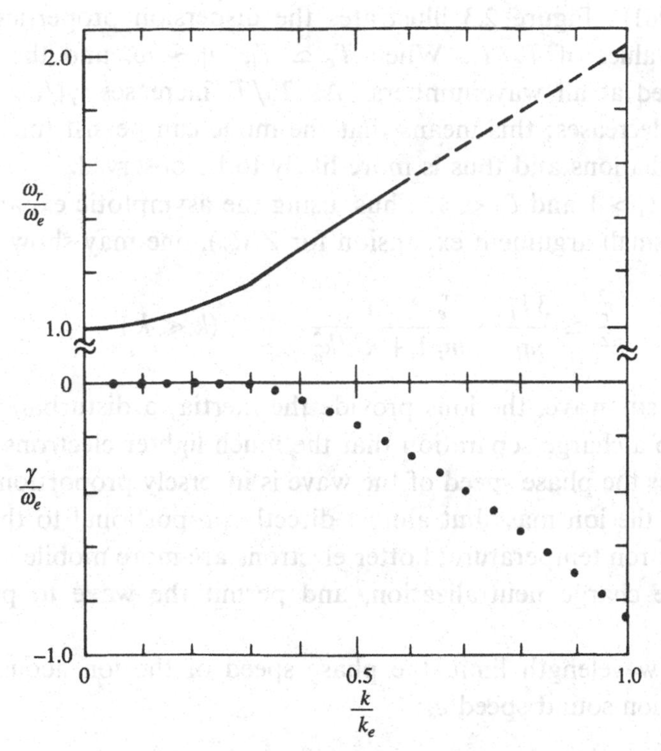

Fig. 2.2 The real frequency (solid/dashed line) and damping rate (dotted line) of an electron plasma wave as functions of wavenumber. A dashed line indicates that the mode is heavily damped ($\gamma < -|\omega_r|/2\pi$). Here, and in all following figures of Section 2.2, $\mathbf{B}_o = 0$.

yields the standard results

$$\omega_r^2 = \omega_e^2 + 3k^2 v_e^2 \qquad (k \ll k_e), \tag{2.2.13}$$

$$\gamma = -\left(\frac{\pi}{8}\right)^{1/2} \frac{k_e^3}{k^3} \frac{\omega_r^2}{\omega_e} \exp(-\omega_r^2/2k^2 v_e^2).$$

Thus, Landau damping of this wave, which is also called the Langmuir mode, is essentially negligible at relatively long wavelengths. As k/k_e increases, ζ_e becomes smaller, the wave-particle interaction becomes stronger and Landau damping increases, so that near $k \sim k_e$, $\gamma \sim \omega_r$ and the wave no longer propagates.

2.2.2 Ion acoustic waves

If $T_e \gg T_i$, the ion acoustic wave at $\omega_r \lesssim \omega_i$ is the second relatively weakly damped mode that can exist in a collisionless, unmagnetized plasma (Fried and Gould, 1961). Figure 2.3 illustrates the dispersion properties of this mode at two values of T_e/T_i. When $T_e \simeq T_i$, $|\gamma| \sim \omega_r$ and the mode is strongly damped at all wavenumbers. As T_e/T_i increases $|\gamma|/\omega_r$ at small wavenumbers decreases; this means that the mode can persist for a greater number of oscillations and thus is more likely to be observed.

At $T_e \gg T_i$, $\zeta_i > 1$ and $\zeta_e \ll 1$. Thus, using the asymptotic expansion for $Z'(\zeta_i)$ and the small argument expansion for $Z'(\zeta_e)$, one may show that

$$\frac{\omega_r^2}{k^2} \simeq \frac{3T_i}{m_i} + \frac{T_e}{m_i}\frac{1}{1+k^2/k_e^2} \qquad (k \ll k_i). \tag{2.2.14}$$

In an ion acoustic wave, the ions provide the inertia; a disturbance of this species leads to a charge separation that the much lighter electrons move to neutralize. Thus the phase speed of the wave is inversely proportional to the square root of the ion mass but almost directly proportional to the square root of the electron temperature; hotter electrons are more mobile, can more readily provide charge neutralization, and permit the wave to propagate more rapidly.

In the long wavelength limit, the phase speed of the ion acoustic wave reduces to the ion sound speed c_s:

$$c_s \equiv \left(\frac{T_e + 3T_i}{m_i}\right)^{1/2}. \tag{2.2.15}$$

Fluid calculations of ion acoustic phase speeds typically yield $c_s^2 = (\gamma_e T_e + \gamma_i T_i)/m_i$, where γ_j represents the ratio of specific heats of the jth species (Boyd and Sanderson, 1969, Chapter 8). Derivation of the ion sound speed from the Vlasov equation directly demonstrates that the electron and ion responses to the fluctuating electric fields of the ion acoustic wave are isothermal ($\gamma_e = 1$) and one-dimensional adiabatic ($\gamma_i = 3$), respectively. As Figure 2.3 shows, $\omega_r/k \simeq c_s$ is appropriate down to $T_e \simeq T_i$. Although the $3T_i$ contribution to c_s is often ignored, it should be kept for the description of ion acoustic waves and instabilities in any plasma with $T_e \sim T_i$.

In principle, there are an infinite number of solutions to Equation (2.2.8) with (2.2.12). Numerical solutions of the linear dispersion equation do, in fact, yield a series of successively higher frequency modes at a given wavenumber in both the high frequency ($\omega_r < \omega_e$) and low frequency ($\omega_r < \omega_i$) regimes (Fried and Gould, 1961). These modes are typically acoustic-like (i.e. $\omega_r/k \simeq$ constant) and are generally heavily damped ($|\gamma| \sim \omega_r$). Because of this latter property they are not important in plasmas with each species described by a single Maxwellian-like distribution. However, as we

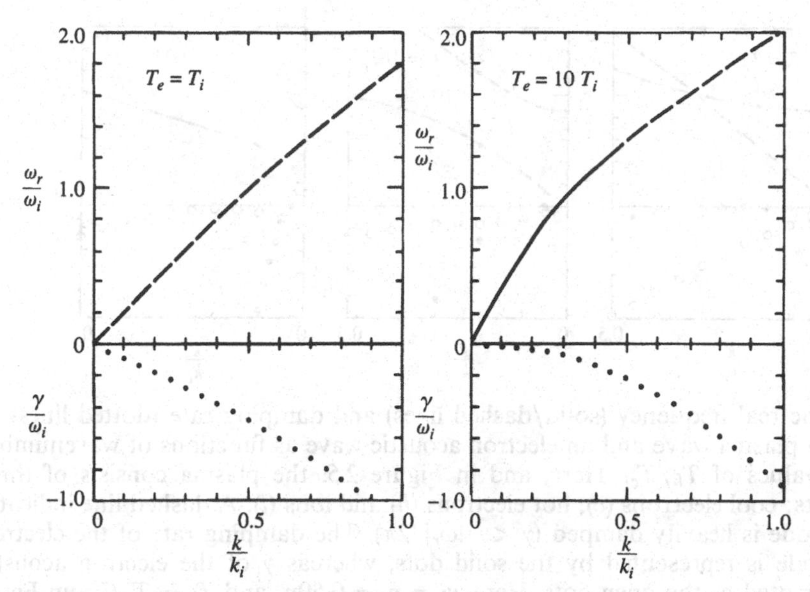

Fig. 2.3 The real frequency (solid/dashed lines) and damping rate (dotted lines) of an ion acoustic wave in an electron-proton plasma as functions of wavenumber for two values of T_e/T_i. A dashed line indicates that the mode is heavily damped ($\gamma < -|\omega_r|/2\pi$). Here, and in all following figures of Chapter 2, $m_i = 1836m_e$.

demonstrate in the next section, at least one of these modes can become lightly damped in the presence of non-Maxwellian distributions.

Problem 2.2.3. Use Figure 2.3 to confirm that, for an ion acoustic wave at $T_e \gg T_i$, $\zeta_i > 1$ and $\zeta_e \ll 1$. Then use Equation (2.2.8) with the susceptibility of Equation (2.2.12) in an electron-ion plasma to derive the phase speed of this wave, Equation (2.2.14). Finally, determine the condition such that ion acoustic waves at $T_e \gg T_i$ correspond to quasineutral density fluctuations ($n_e^{(1)} \simeq n_i^{(1)}$).

2.2.3 Two electron components: ion acoustic and electron acoustic waves

In the presence of non-Maxwellian distributions, the properties of plasma modes may change significantly. Consider a two-component electron distribution function represented as the sum of two Maxwellian components, one hot (denoted by the subscript *h*), the other relatively cool (subscript *c*).

As long as the temperature conditions are not extreme, both electron components remain resonant with the ion acoustic wave ($|\zeta_c| \ll 1$ and $|\zeta_h| \ll 1$) and equations (2.2.14) and (2.2.15) remain valid as long as T_e is

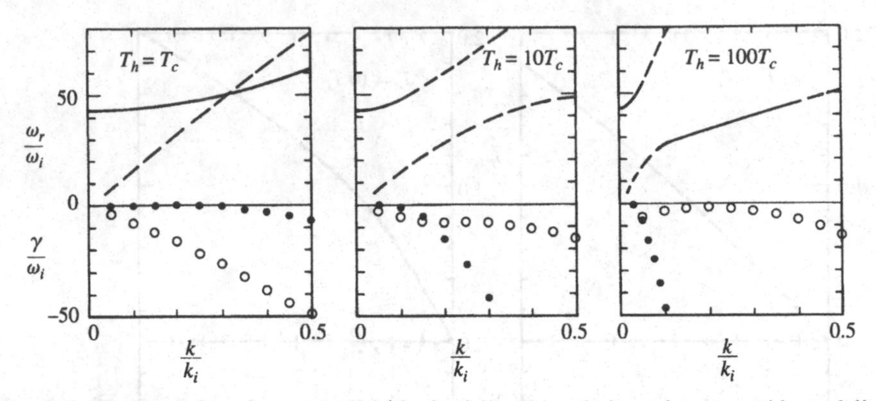

Fig. 2.4 The real frequency (solid/dashed lines) and damping rate (dotted lines) of an electron plasma wave and an electron acoustic wave as functions of wavenumber for three values of T_h/T_c. Here, and in Figure 2.5, the plasma consists of three components: cool electrons (*c*), hot electrons (*h*) and ions (*i*). A dashed line indicates that the mode is heavily damped ($\gamma < -|\omega_r|/2\pi$). The damping rate of the electron plasma mode is represented by the solid dots, whereas γ of the electron acoustic mode is denoted by the open dots. Here, $n_h = n_c = 0.50n_e$ and $T_c = T_i$ (From Fig. 1 of Gary and Tokar, 1985).

replaced by an effective temperature defined as (Jones *et al.*, 1975)

$$T_{eff} \equiv n_e \frac{T_c T_h}{n_h T_c + n_c T_h}. \tag{2.2.16}$$

In addition, if $n_h \sim n_c$ and T_h/T_c is sufficiently large, another acoustic-like mode can become lightly damped. Figure 2.4 illustrates the evolution of this acoustic-like mode as the electron component temperature ratio is increased. At $T_h = T_c$, the weakly damped electron plasma mode and a strongly damped acoustic-like mode are shown. As T_h/T_c is increased, the phase speed of the electron plasma mode increases, as does its damping at fixed k/k_i. In contrast, the damping of the acoustic-like mode decreases at intermediate wavenumbers so that, by $T_h = 100T_c$, it has become relatively lightly damped ($|\gamma| < \omega_r/2\pi$).

We call this wave the "electron acoustic mode" (Watanabe and Taniuti, 1977). As the rightmost panel of Figure 2.4 shows (Gary and Tokar, 1985), there are actually three regimes for this mode at $T_h \gg T_c$. At small wavenumbers there is the heavily damped acoustic regime with $\omega_r^2/k^2 \simeq (n_c/n_h)v_h^2$. At intermediate wavenumbers $k_h < k < k_c$ there is a lightly damped regime near the plasma frequency of the cool electron component with

$$\omega_r^2 \simeq \omega_c^2 \frac{1 + 3k^2/k_c^2}{1 + k_h^2/k^2}. \tag{2.2.17}$$

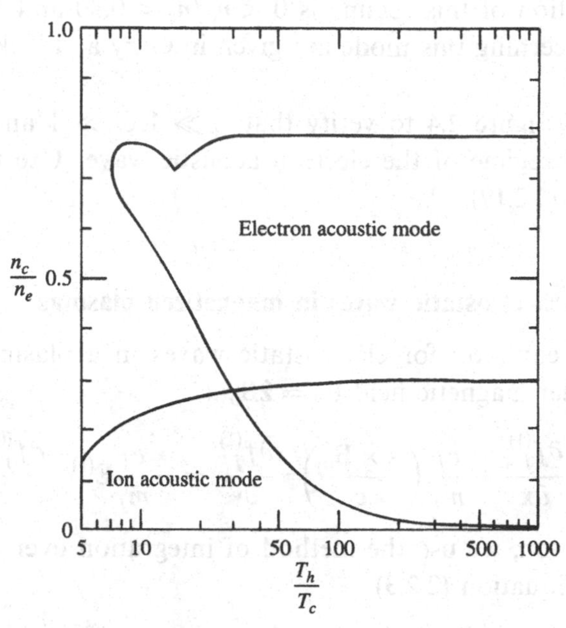

Fig. 2.5 The lightly damped regime ($|\gamma| \leq |\omega_r|/2\pi$) of the electron acoustic mode (acoustic-like dispersion at small wavenumbers and $\omega_i < \omega_r < \omega_e$) and the ion acoustic wave (ion acoustic dispersion and $\omega_r < \omega_i$) in the parameter space of cool plasma density vs. hot-to-cool electron component temperature ratio. Here $T_c = T_i$ (From Fig. 4 of Gary and Tokar, 1985).

Finally there is another heavily damped regime at higher frequencies and shorter wavelengths. Because $\omega_r \gg \omega_i$ in the lightly damped regime, the ions are nonresonant ($\zeta_i \gg 1$) and do not make a significant contribution to the dispersion equation. At intermediate wavelengths, $\zeta_c \gg 1$ and $\zeta_h \ll 1$, so that neither electron component contributes significant Landau damping. The cool electrons are the less mobile electron component and provide the wave inertia; thus their role is analogous to that of the ions in an ion acoustic wave. The hot electrons are more mobile and act as the fluid that flows to neutralize the charge separation created by the cool electron oscillations; by analogy with the ion acoustic wave, the phase speed of this mode at long wavelengths is proportional to the square root of the temperature of this component. At small k, the phase speed and ζ_h both increase so that the hot component leads to strong damping; conversely, at large wavenumber, ω_r/k decreases sufficiently so that the cool electrons become resonant and provide the damping in that regime.

The parameter regime in which the electron acoustic wave is a normal mode of the plasma depends upon the relative densities and temperatures of the two components. This is illustrated in Figure 2.5, which shows that a crude approximation of this regime is $0 < n_c/n_e < 0.80$ and $10 < T_h/T_c$. Further details concerning this mode are given in Gary and Tokar (1985).

Problem 2.2.4. Use Figure 2.4 to verify that $\zeta_i \gg 1, \zeta_c \gg 1$ and $\zeta_h \ll 1$ in the lightly damped regime of the electron acoustic wave. Use these results to derive Equation (2.2.17).

2.3 Electrostatic waves in magnetized plasmas

The linear Vlasov equation for electrostatic waves in a plasma bearing a uniform zeroth-order magnetic field $\mathbf{B}_o = \hat{\mathbf{z}}B_o$ is

$$\frac{\partial f_j^{(1)}(\mathbf{x},\mathbf{v},t)}{\partial t} + \mathbf{v}\cdot\frac{\partial f_j^{(1)}}{\partial \mathbf{x}} + \frac{e_j}{m_j}\left(\frac{\mathbf{v}\times\mathbf{B}_o}{c}\right)\cdot\frac{\partial f_j^{(1)}}{\partial \mathbf{v}} = -\frac{e_j}{m_j}\mathbf{E}^{(1)}\cdot\frac{\partial f_j^{(0)}(\mathbf{v})}{\partial \mathbf{v}}. \quad (2.3.1)$$

To solve this equation, we use the method of integration over unperturbed orbits. Thus with Equation (2.2.3)

$$f_j^{(1)}(\mathbf{x},\mathbf{v},t) = \frac{e_j}{m_j}\int_{-\infty}^{t} dt'\ \nabla'\phi^{(1)}(\mathbf{x}',t')\cdot\frac{\partial f_j^{(0)}(\mathbf{v}')}{\partial \mathbf{v}'} \quad (2.3.2)$$

where the primed variables denote the unperturbed orbit of a charged particle in $\mathbf{B}_o$. Assuming that all first-order quantities may be written as Equation (2.2.2), we obtain

$$f_j^{(1)}(\mathbf{k},\mathbf{v},\omega) = \frac{e_j}{m_j}\phi^{(1)}(\mathbf{k},\omega)\int_{-\infty}^{t} dt'\ i\mathbf{k}\cdot\frac{\partial f_j^{(0)}(\mathbf{v}')}{\partial \mathbf{v}'} \exp\{i[\mathbf{k}\cdot(\mathbf{x}'-\mathbf{x})-\omega(t'-t)]\}. \quad (2.3.3)$$

Integrating over velocity to obtain the fluctuating density yields

$$n_j^{(1)}(\mathbf{k},\omega) = \frac{e_j}{m_j}\phi^{(1)}(\mathbf{k},\omega)\int d^3v\int_{-\infty}^{t} dt'\ i\mathbf{k}\cdot\frac{\partial f_j^{(0)}(\mathbf{v}')}{\partial \mathbf{v}'} \exp\{i[\mathbf{k}\cdot(\mathbf{x}'-\mathbf{x})-\omega(t'-t)]\}. \quad (2.3.4)$$

Using Poisson's equation and the definition of the jth component susceptibility K_j from Equation (2.2.5) the linear dispersion equation is Equation (2.2.8) with

$$K_j(\mathbf{k},\omega) = -\frac{\omega_j^2}{n_j k^2}\int d^3v\int dt'\ i\mathbf{k}\cdot\frac{\partial f_j^{(0)}}{\partial \mathbf{v}'}\exp\{i[\mathbf{k}\cdot(\mathbf{x}'-\mathbf{x})-\omega(t'-t)]\}. \quad (2.3.5)$$

If we now assume that the zeroth-order distributions are Maxwellian, we use

$$\frac{\partial}{\partial t'} \exp\left\{i\left[\mathbf{k}\cdot(\mathbf{x}'-\mathbf{x})-\omega(t'-t)\right]\right\} = i(\mathbf{k}\cdot\mathbf{v}'-\omega) \exp\left\{i\left[\mathbf{k}\cdot(\mathbf{x}'-\mathbf{x})-\omega(t'-t)\right]\right\} \tag{2.3.6}$$

to obtain

$$K_j^{(M)}(\mathbf{k},\omega) = \frac{k_j^2}{k^2}\left\{1+\frac{i\omega}{n_j}\int d^3v f_j^{(M)}(v)\int_{-\infty}^{0} d\tau \exp\left[ib_j(\tau,\omega)\right]\right\} \tag{2.3.7}$$

where $b_j(\tau,\omega) = \mathbf{k}\cdot(\mathbf{x}'-\mathbf{x})-\omega\tau, \tau \equiv t'-t$ and the superscript (M) denotes a quantity associated with a Maxwellian distribution. Here we assume a uniform field $\mathbf{B}_o = \hat{\mathbf{z}}B_o$ so that, from Appendix B,

$$b_j(\tau,\omega) = \frac{k_y v_x}{\Omega_j}(\cos\Omega_j\tau - 1) + \frac{k_y v_y}{\Omega_j}\sin\Omega_j\tau + (k_z v_z - \omega)\tau \tag{2.3.8a}$$

$$= \frac{k_y v_\perp}{\Omega_j}[\cos(\Omega_j\tau-\phi)-\cos\phi] + (k_z v_z-\omega)\tau. \tag{2.3.8b}$$

The above integral is a standard form that frequently arises in linear Vlasov theory. Its evaluation is detailed in Appendix C; from Equation (C.6) the susceptibility is

$$K_j^{(M)}(\mathbf{k},\omega) = \frac{k_j^2}{k^2}\left[1+\frac{\omega\exp(-\lambda_j)}{\sqrt{2}|k_z|v_j}\sum_{m=-\infty}^{\infty} I_m(\lambda_j)Z(\zeta_j^m)\right] \tag{2.3.9}$$

where $\lambda_j \equiv (k_y a_j)^2$ and $\zeta_j^m \equiv (\omega + m\Omega_j)/\sqrt{2}|k_z|v_j$.

The presence of the magnetic field has generalized the resonance condition discussed in Section 2.2. From Equation (C.3) it is clear that a resonant wave-particle interaction will arise for the jth component whenever

$$\omega = k_z v_z - m\Omega_j \qquad (m = 0, \pm 1, \pm 2, ...). \tag{2.3.10}$$

The Landau resonance at $m = 0$ is modified so that it now concerns only the velocity component parallel to $\mathbf{B}_o$. And $m \neq 0$ corresponds to a different wave-particle interaction, the cyclotron resonance. Particles satisfying this condition move such that $\mathbf{E}^{(1)}\cdot\mathbf{v}_\perp$ does not change sign so that there is a strong exchange of energy between the wave and the perpendicular motion of those particles.

In the limit $k_y = 0, k_z = k$ and Equation (2.3.9) reduces to Equation (2.2.12). Thus at $\mathbf{k}\times\mathbf{B}_o = 0$, linear dispersion reduces to that of an unmagnetized plasma and the results of Section 2.2 are recovered.

In the limit of perpendicular propagation, $k_z = 0$, and by the asymptotic expansion of the plasma dispersion function,

$$K_j^{(M)}(\mathbf{k}, \omega) = \frac{k_j^2}{k^2}\left\{1 - \omega \exp(-\lambda_j) \sum_{m=-\infty}^{\infty} \frac{I_m(\lambda_j)}{\omega + m\Omega_j}\right\}. \tag{2.3.11}$$

Problem 2.3.1. Construct an argument to show that the cyclotron resonance at

$$v_z = \frac{\omega_r \pm \Omega_j}{k_z}$$

corresponds to strong energy exchange between an obliquely propagating electrostatic wave and the cyclotron motion of a charged particle. (Hint: work in the frame of the wave where $\omega_r = 0$).

Problem 2.3.2. Use the zero temperature limit in Equation (2.3.9) to show that the cold plasma result for the susceptibility of the jth species is

$$K_j(\mathbf{k}, \omega) = -\omega_j^2 \left(\frac{\sin^2\theta}{\omega^2 - \Omega_j^2} + \frac{\cos^2\theta}{\omega^2}\right)$$

where $\hat{\mathbf{k}} \cdot \hat{\mathbf{B}}_o = \cos\theta$.

2.3.1 Oblique propagation: ion acoustic and ion cyclotron waves

As an example of propagation at arbitrary angles with respect to the magnetic field, we consider solutions to the electrostatic dispersion equation (2.2.8) with Equation (2.3.9) for frequencies near and below the ion cyclotron frequency. Some typical results are illustrated in Figure 2.6, which plots the real part of the frequency, ω_r, and the damping rate, γ, as a function of wavenumber k for three different values of θ.

For $\mathbf{k} \times \mathbf{B}_o = 0$, the ion acoustic mode of Section 2.2.2 is present, with $\omega_r/k \simeq c_s$. But as $\mathbf{k}$ becomes oblique to $\mathbf{B}_o$, the presence of the magnetic field alters the dispersion properties by comparison with the case of an unmagnetized plasma. At $\omega_r \ll \Omega_i$ and $ka_i \ll 1$, the ion acoustic wave now satisfies

$$\frac{\omega_r}{k_z} \simeq c_s. \tag{2.3.12}$$

The primary difference between this equation and the long wavelength limit of Equation (2.2.14) is the extra $\cos\theta$ factor introduced by the k_z here, implying that this mode cannot propagate across $\mathbf{B}_o$. This happens because the magnetic field inhibits cross-field motion of the particles, and the electron

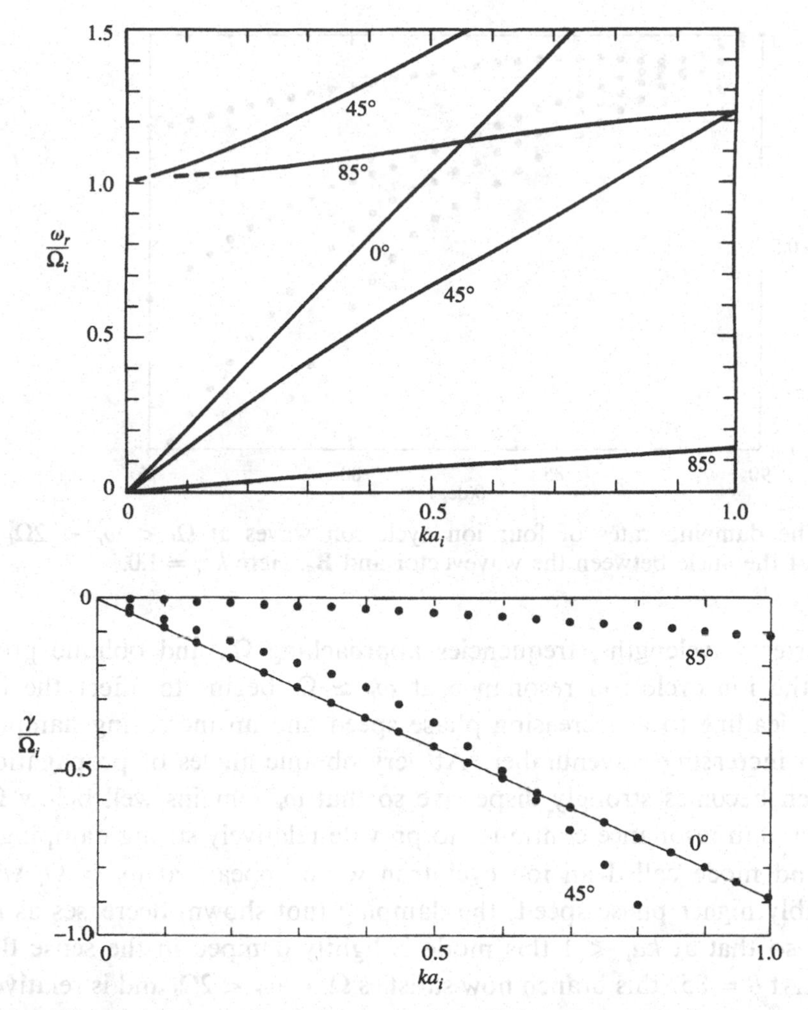

Fig. 2.6 The real frequencies (upper panel) and damping rates (lower panel) of low frequency electrostatic waves as functions of wavenumber for three different values of θ. The solid line in the lower panel connects the damping rates at $\theta = 0°$. Here, as in Figures 2.7 and 2.8, we consider an electron-proton plasma. The damping rates of only the lower ω_r branch are illustrated. Here, and for all the figures of Section 2.3, $\omega_e = 10|\Omega_e|$ and $T_e = T_i$.

response that allows the acoustic mode to propagate can only take place along $\mathbf{B}_o$. In the long wavelength limit, the damping ratio γ/ω_r remains constant as θ increases.

Problem 2.3.3. Use the magnetized susceptibility (2.3.9) for both electrons and ions in the dispersion equation (2.2.8) to derive Equation (2.3.12) in the long wavelength limit.

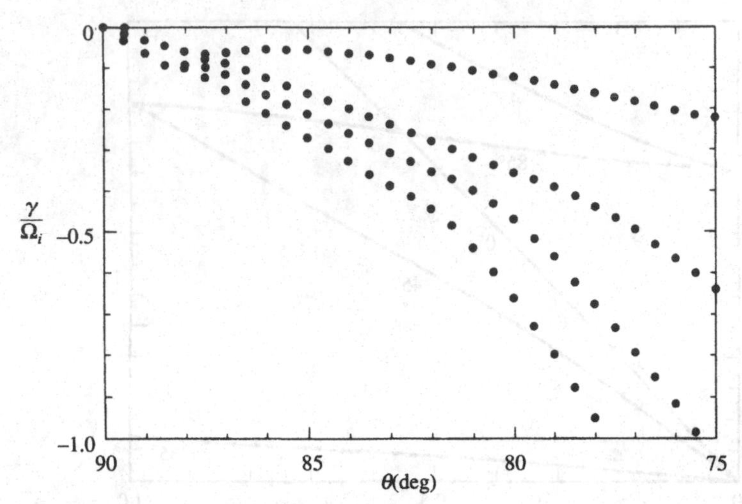

Fig. 2.7 The damping rates of four ion cyclotron waves at $\Omega_i < \omega_r < 2\Omega_i$ as functions of the angle between the wavevector and $\mathbf{B}_o$. Here $ka_i = 1.0$.

At shorter wavelengths, frequencies approaching Ω_i, and oblique propagation, the ion cyclotron resonance at $\omega_r \simeq \Omega_i$ begins to affect the ion dynamics, leading to a decreasing phase speed and an increasing damping ratio with increasing wavenumber. At very oblique angles of propagation, this branch becomes strongly dispersive so that ω_r remains well below Ω_i; the ion Landau resonance continues to provide relatively strong damping.

A second mode called an ion cyclotron wave appears at $\omega_r > \Omega_i$ with considerably higher phase speed; the damping (not shown) decreases as ka_i decreases so that at $ka_i \ll 1$ this mode is lightly damped in the sense that $|\gamma| \ll \omega_r$. At $\theta = 85°$ this branch now satisfies $\Omega_i < \omega_r < 2\Omega_i$ and is relatively lightly damped for all wavenumbers shown.

At higher frequencies, this picture is repeated: the ion acoustic mode at $\theta = 0°$ splits into cyclotron modes that are bounded by successive multiples of the ion cyclotron frequency. At $ka_i \ll 1$ and a given θ, the damping of successively higher cyclotron harmonic modes increases with harmonic number, so that, in general, the lower order modes have a greater range of propagation about the perpendicular and are therefore easier to observe.

However, the ion cyclotron wave depicted in Figure 2.6 is not the only mode present between Ω_i and $2\Omega_i$. Computer solutions of the full dispersion equation show that there are several other modes present in this frequency range. In contrast to the modes in an unmagnetized plasma mentioned in Subsection 2.2.2, these waves may propagate because, at θ sufficiently close to 90°, they satisfy $|\gamma| \ll \omega_r$.

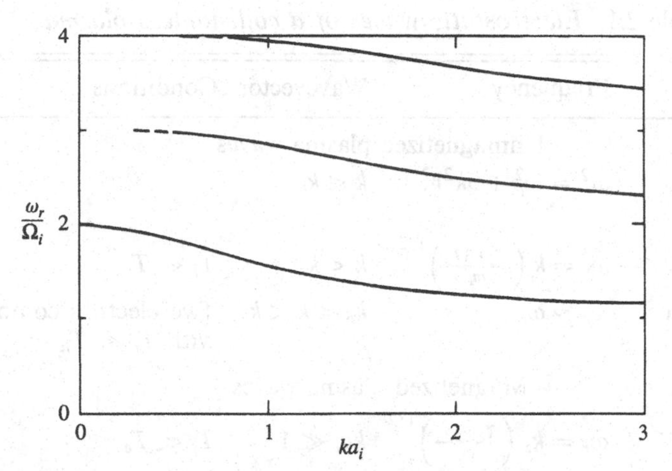

Fig. 2.8 The real frequencies of the first three ion Bernstein modes as functions of wavenumber at $\theta = 90°$.

The damping rates of four ion cyclotron waves at $ka_i = 1$ are illustrated in Figure 2.7. At this wavenumber, the least damped mode at $\theta \leq 89°$ is the ion cyclotron wave of Figure 2.6, with an ω_r that approaches Ω_i as θ goes to 90°. However, the mode of Figure 2.7 that is most heavily damped at smaller θ becomes the least damped for $89° < \theta \leq 90°$. This wave does not approach Ω_i as k_z goes to zero and is the only mode of the four shown that actually survives to perpendicular propagation. The special limiting case of $\theta = 90°$ is discussed in the next section.

2.3.2 Perpendicular propagation: ion Bernstein modes

The linear dispersion properties of waves that propagate perpendicular to the magnetic field have several special properties. If Equation (2.3.11) is used in the dispersion Equation (2.2.8), it follows that there is no damping. If one assumes ω is real, then K_j is also real and $\gamma = 0$ is a consistent solution of the dispersion equation. Because the wave particle resonance condition (2.3.10) cannot be satisfied by any finite v_z in the limit of vanishing k_z, there is neither Landau nor cyclotron damping.

As in the case of oblique propagation, a series of cyclotron harmonic waves exist, satisfying

$$m\Omega_i < |\omega_r| < (m+1)\Omega_i \qquad (m = 1, 2, 3, ...).$$

These waves are called "ion Bernstein modes," after their discoverer (Bernstein, 1958). Unlike the oblique case, however, there is only one mode for each value of m. The $m = 1$, 2, and 3 waves are shown in Figure 2.8. A

Table 2.1. *Electrostatic waves of a collisionless plasma*

Name	Frequency	Wavevector	Conditions
	Unmagnetized plasma waves		
Electron plasma (Langmuir)	$\omega_r^2 = \omega_e^2 + 3k^2v_e^2$	$k < k_e$	
Ion acoustic	$\omega_r = k\left(\frac{T_e+3T_i}{m_i}\right)^{1/2}$	$k < k_e$	$T_i \ll T_e$
Electron acoustic	$\omega_r \sim \omega_c$	$k_h < k < k_c$	Two electron components with $T_c \ll T_h$
	Magnetized plasma waves		
Ion acoustic	$\omega_r = k_z\left(\frac{T_e+3T_i}{m_i}\right)^{1/2}$	$ka_i \ll 1$	$T_i \ll T_e$
Ion Bernstein (Ion cyclotron)	$\omega_r \simeq m\Omega_i$ $m = \pm 1, \pm 2, ...$	$ka_i \sim 1$	$0 \leq k_z \ll k_y$
Electron Bernstein (Electron cyclotron)	$\omega_r \simeq m\Omega_e$ $m = \pm 1, \pm 2, ...$	$ka_e \sim 1$	$0 \leq k_z \ll k_y$

similar set of electron Bernstein modes exist at perpendicular propagation between harmonics of the electron gyrofrequency (Stone and Auer, 1965).

2.4 Summary

The lightly damped electrostatic modes of a collisionless plasma that have been described in this chapter are summarized in Table 2.1. The upper three modes in this table are independent of the magnetic field and propagate in an unmagnetized plasma as well as preferentially along the magnetic field when $\mathbf{B}_o \neq 0$. The lower three modes propagate only in magnetized plasmas.

Non-Maxwellian distribution functions may cause many of these modes to become unstable. However, the total number of electrostatic instabilities is considerably greater than the number of normal modes, because several different free energy sources can cause the same mode to become unstable, and because some free energy sources can give rise to growing modes that do not propagate in a Maxwellian plasma. These points are illustrated by the discussions of Chapter 3, which treat electrostatic instabilities in collisionless plasmas.

References

Bernstein, I. B., Waves in a plasma in a magnetic field, *Phys. Rev.*, **109**, 10, 1958.

Boyd, T. J. M., and J. J. Sanderson, *Plasma Dynamics*, Barnes and Noble, New York, 1969.

Fried, B. D., and R. W. Gould, Longitudinal ion oscillations in a hot plasma, *Phys. Fluids*, **4**, 139, 1961.

Gary, S. P., and R. L. Tokar, The electron acoustic mode, *Phys. Fluids*, **28**, 2439, 1985.

Jones, W. D., A. Lee, S. M. Gleman and H. J. Doucet, Propagation of ion-acoustic waves in a two-electron-temperature plasma, *Phys. Res. Lett.*, **35**, 1349, 1975.

Landau, L., On the vibrations of the electronic plasma, *J. Phys. (USSR)*, **10**, 25, 1946.

Montgomery, D. C., *Theory of the Unmagnetized Plasma*, Gordon and Breach, New York, 1971.

Montgomery, D. C., and D. A. Tidman, *Plasma Kinetic Theory*, McGraw-Hill, New York, 1964.

Stone, P. M., and P. L. Auer, Excitation of electrostatic waves near electron cyclotron harmonic frequencies, *Phys. Rev.*, **138**, A695, 1965.

Watanabe, K., and T. Taniuti, Electron-acoustic mode in a plasma of two-temperature electrons, *J. Phys. Soc. Japan*, **43**, 1819, 1977.

3

Electrostatic component/component instabilities in uniform plasmas

This chapter begins our consideration of instabilities, plasma modes that grow in time or space. The source of growth of a plasma microinstability is what is imprecisely called "free energy:" an anisotropy or inhomogeneity in the zeroth-order velocity distribution function. In this chapter we consider free energy sources associated with the relative drifts of plasma components, and find that different types of relative drifts each can give rise to several different unstable modes.

As in Chapter 2, we consider uniform collisionless plasmas in which the evolution of the distribution function of the jth component is described by the Vlasov equation (1.3.2). Again we restrict ourselves to electrostatic fluctuations; that is, we assume there are no fluctuating magnetic fields and the fluctuating electric fields are derived from Poisson's equation (1.2.4).

In Section 3.1 we state zeroth-order distribution functions representing several different free energy sources. Section 3.2 considers electrostatic instabilities driven by component/component relative drifts in unmagnetized plasmas and Section 3.3 considers electrostatic instabilities driven by the same free energy sources in magnetized plasmas. Throughout this chapter, we assume the plasma to be charge neutral and to bear no steady-state electric field.

Although virtually all space plasmas bear ambient magnetic fields, some waves and instabilities have properties that are essentially independent of $\mathbf{B}_o$. In Section 3.3 we demonstrate this by showing that the magnetized electrostatic dispersion equation at $\mathbf{k} \times \mathbf{B}_o = 0$ reduces to the unmagnetized form. More generally, we also show that, at arbitrary directions of propagation, a jth component has an unmagnetized response to fluctuating electrostatic fields if $|\Omega_j| \ll \gamma$. Thus it is frequently appropriate to discuss the application of unmagnetized electrostatic instabilities to magnetized space plasmas, and we do so in Section 3.2. Some space plasma applications of magnetized electrostatic instabilities are discussed in Section 3.3, and Section 3.4 is a brief summary.

3.1 Zeroth-order distributions: free energy sources

The zeroth-order distribution functions used in linear Vlasov theory must satisfy the zeroth-order Vlasov equation. In a field-free, uniform plasma, any function of v_x, v_y and v_z satisfies this criterion. If, however, a uniform zeroth-order magnetic field $\mathbf{B}_o = \hat{\mathbf{z}}B_o$ is present, the time-independent, zeroth-order Vlasov equation is

$$\left(\frac{\mathbf{v} \times \mathbf{B}_o}{c}\right) \cdot \frac{\partial f_j^{(0)}}{\partial \mathbf{v}} = 0, \tag{3.1.1}$$

which implies that any zeroth-order distribution must satisfy

$$f_j^{(0)}(\mathbf{v}) = f_j^{(0)}(v_z, v_\perp^2). \tag{3.1.2}$$

Problem 3.1.1. Prove that the zeroth-order distribution function that satisfies Equation (3.1.1) is independent of ϕ, the azimuthal coordinate in cylindrical velocity coordinates.

A common source of free energy in a collisionless plasma is an electric current. For example, if a uniform electric field penetrates an unmagnetized collisionless plasma, each charged particle will experience an acceleration independent of its velocity. Electrons and ions will move in opposite directions and a net current $\mathbf{J}_o$ will be established. But if the species velocity distributions are Maxwellian to begin with, they will remain Maxwellian in their accelerated frames of reference. If the electric field is sufficiently weak that the acceleration rate is slow compared to the linear growth rate, it is a good approximation to write the zeroth-order distributions of the two species as drifting Maxwellians:

$$f_j^{(0)}(\mathbf{v}) = \frac{n_j}{(2\pi v_j^2)^{3/2}} \exp\left[-\frac{(\mathbf{v} - \mathbf{v}_{oj})^2}{2v_j^2}\right]. \tag{3.1.3}$$

In a magnetized plasma, Equation (3.1.3) is of the form (3.1.2) only if the drift velocity is parallel or antiparallel to $\mathbf{B}_o$. In physical terms, a drift across a uniform magnetic field cannot be maintained in a homogeneous plasma if no electric field is present. If there is an $\mathbf{E}_o$ perpendicular to $\mathbf{B}_o$, all charged species will undergo the same $\mathbf{E} \times \mathbf{B}$ drift. In this case there would be no relative drifts between species or components, and there would be no instabilities. Since we limit our discussions to homogeneous or, as in Chapter

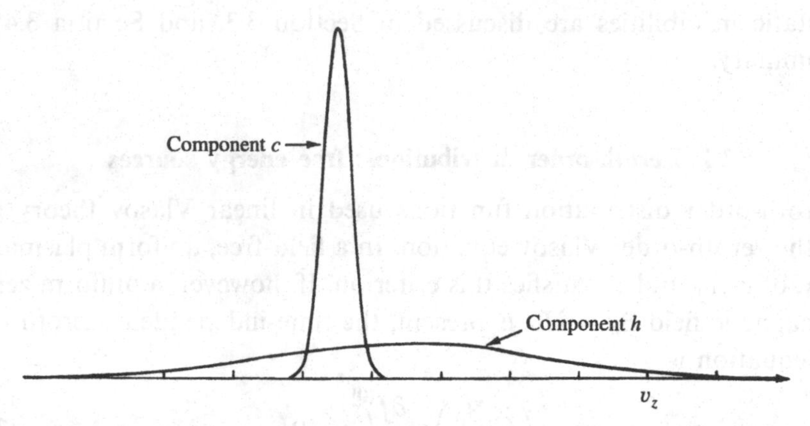

Fig. 3.1 The reduced distribution functions for two Maxwellian components (c and h) with relative thermal speeds $v_h/v_c = 100$ and relative drift speed $v_o = 5v_c$. If $m_c = m_h$, c is the cool component and h is the hot component; if $T_c = T_h$, c is the heavy component and h is the light component.

4, weakly inhomogeneous plasmas, we do not consider such cross-field drifts and assume $\mathbf{v}_{oj} = \hat{\mathbf{z}}v_{oj}$ throughout both this chapter and Chapter 8.

The reduced distribution functions for a modest electron/ion drift, $T_e = T_i$ and $m_i = 100m_e$, are shown in Figure 3.1. A more realistic value of m_e/m_i will make the ion distribution appear much more narrow relative to the $f_e(v_z)$.

There are many other zeroth-order distributions which can lead to instability growth even if $\mathbf{J}_o = 0$. One example of such a free energy is the relative streaming of two ion components against an electron background such that the three drift speeds satisfy $\sum_j e_j n_j v_{oj} = 0$. We use the term "ion/ion" to describe such instabilities; similarly, we will employ "electron/electron" to describe modes driven unstable by the relative streaming of two electron components.

We use the term "component" to denote a subset of electrons or ions that is readily distinguishable in velocity space. If each of these is Maxwellian-like, then Equation (3.1.3) can be used to describe the zeroth-order distributions of each component. For example, a two-component species can be created by the injection of a relatively tenuous, energetic particle population into a more dense, thermal plasma. The notation we use in this case is a subscript b to denote the more tenuous beam and c to designate the more dense core component; thus, $n_b \ll n_c$. The relative drift between two components of the same species is denoted by

$$\mathbf{v}_o = \mathbf{v}_{ob} - \mathbf{v}_{oc}.$$

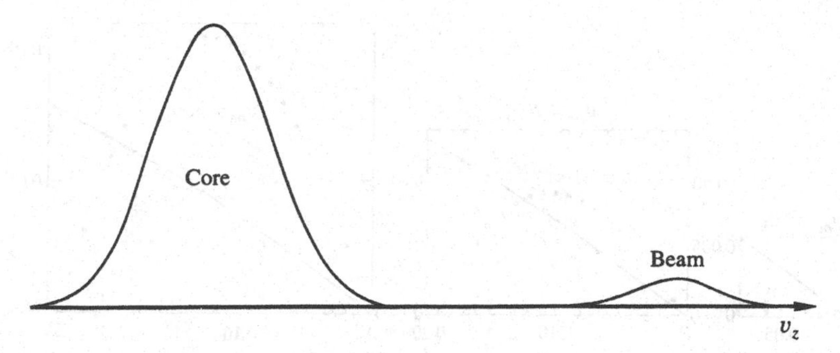

Fig. 3.2 A beam/core reduced distribution function.

A representative beam/core distribution is shown in Figure 3.2.

Another example we use concerns two components of the same species and similar densities but significantly different temperatures. In this case, as in Subsection 2.2.3, we denote the hot component by subscript h and the cool component by subscript c; the relative drift is

$$\mathbf{v}_o = \mathbf{v}_{oh} - \mathbf{v}_{oc}.$$

The reduced distribution illustrated in Figure 3.1 depicts this configuration, as well as the electron/ion case.

3.2 Electrostatic component/component instabilities in unmagnetized plasmas

The derivation of the linear dispersion equation in this case proceeds just as in Section 2.2. Thus, from Equation (2.2.8),

$$1 + \sum_j K_j(\mathbf{k}, \omega) = 0 \tag{3.2.1}$$

where the jth component susceptibility is Equation (2.2.7)

$$K_j(\mathbf{k}, \omega) = -\frac{\omega_j^2}{n_j}\frac{\partial}{\partial \omega}\int \frac{d^3v f_j^{(0)}(\mathbf{v})}{\mathbf{k}\cdot\mathbf{v} - \omega}. \tag{3.2.2}$$

Throughout this section, we assume that the zeroth-order distribution of each component is a drifting Maxwellian. Using Equation (3.1.3), the susceptibility has the same form as Equation (2.2.12):

$$K_j(\mathbf{k}, \omega) = -\frac{k_j^2}{2k^2} Z'(\zeta_j), \tag{3.2.3}$$

but the argument of the plasma dispersion function is now $\zeta_j = (\omega - \mathbf{k}\cdot\mathbf{v}_{oj})/\sqrt{2}kv_j$.

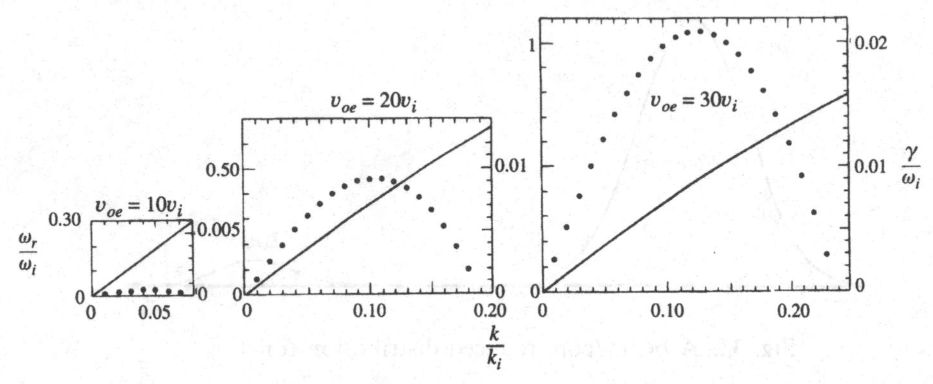

Fig. 3.3 The real frequency (solid lines) and growth rate (dotted lines) of the electron/ion acoustic instability as functions of wavenumber for three different values of the electron/ion relative drift speed. Here and in Figures 3.4 through 3.6, we consider an electron-proton plasma. Here $T_e = 10T_i$ and $v_{oi} = 0$. Here, and in all subsequent figures of Chapter 3, $m_i = m_p = 1836m_e$.

3.2.1 Electron/ion instabilities

In Chapter 2 we examined the waves that may propagate in a homogeneous electron-ion plasma described by Maxwellian distribution functions. One of the simplest ways to perturb such an equilibrium configuration, and thereby introduce a free energy capable of driving instabilities, is to introduce a relative drift between the electrons and ions. We examine such instabilities in this section by considering a two-species plasma with drifting electrons and stationary ions ($v_{oi} = 0$), so that there is a nonzero current in the plasma. If Equation (3.2.1) with (3.2.3) is solved in this configuration, there are no important changes in the dispersion of the electron plasma wave in the electron frame, but the ion acoustic mode becomes unstable for sufficiently large v_{oe} (Fried and Gould, 1961). We call this mode, and more generally any ion acoustic mode driven unstable by an electron component drifting relative to an ion component, the "electron/ion acoustic instability."

Typical results are shown in Figure 3.3. At $v_{oe} \ll v_e$, the electron/ion drift makes little difference in the real frequency of the ion acoustic wave; Equation (2.2.14) is a reasonably good approximation in this regime. As in the zero drift case, the electrons are Landau resonant ($|\zeta_e| \ll 1$) and, if $T_i \ll T_e$, the ions are nonresonant ($|\zeta_i| \gg 1$) with this mode. However, as v_{oe} is increased from zero, the damping of the wave of Figure 2.3 decreases until the instability threshold is reached. Just above this threshold, the electron/ion acoustic instability arises with growth only at relatively long wavelengths. As v_{oe} increases further, γ_m shifts to shorter wavelengths and $\gamma > 0$ over the full range of wavenumbers $0 < k \lesssim k_e$.

If one uses the small and large argument expansions of Appendix A for $Z'(\zeta_e)$ and $Z'(\zeta_i)$, respectively, in Equation (3.2.3), an analytic reduction of the dispersion Equation (3.2.1) may be obtained for the electron/ion acoustic instability. Assuming $v_{oe} \ll v_e$, $T_i \ll T_e$ and $|\gamma| \ll \omega_r$, the real part of Equation (3.2.1) yields ω_r as in the stable case (Equation (2.2.14)) while the imaginary part of the dispersion equation implies

$$\gamma = \frac{\omega_r^3 \pi}{2k^3}\left(\frac{m_i}{T_e}\right)\frac{(\mathbf{k}\cdot\mathbf{v}_{oe} - \omega_r)}{(2\pi v_e^2)^{1/2}}\exp[-(\omega_r/k_z - v_{oe})^2/2v_e^2]. \tag{3.2.4}$$

This expression clearly shows that the electron/ion drift speed must exceed the threshold value of c_s in order that the mode become unstable. Ion Landau damping, which is ignored in this expression, becomes important as T_e/T_i decreases and, by contributing a negative term to the right hand side, raises the drift speed threshold.

Problem 3.2.1. Show that, at $k_z \ll k_e$, the threshold of the electron/ion acoustic instability occurs at

$$v_{oe} \simeq \frac{\omega_r}{k_z}\left[1 + \left(\frac{T_e}{T_i}\right)^{3/2}\left(\frac{m_i}{m_e}\right)^{1/2}\exp(-3/2 - T_e/2T_i)\right].$$

Give a brief physical interpretation of the two right-hand terms.

Thus the threshold drift speed of the electron/ion acoustic instability is a sensitive function of T_e/T_i. Figure 3.4 illustrates this dependence. If one plots v_{oe}/v_e at threshold vs T_e/T_i, the associated curve decreases monotonically as the temperature ratio increases (Fried and Gould, 1961). Here, however, we have normalized the threshold drift with respect to the ion thermal speed; then at $T_e/T_i \lesssim 0.60$ and $T_e/T_i \gtrsim 15.0$ (where threshold corresponds to $k > 0$) v_{oe}/v_i at threshold actually increases with T_e/T_i.

At $v_{oe} \gtrsim v_e$, the wavenumber at maximum growth begins to decrease, ω_r begins to show appreciable change from Equation (2.2.14) and both electrons and ions become nonresonant ($1 \ll |\zeta_j|$ with $j = e, i$). Here the instability enters the cold plasma regime, where it is known as the Buneman or electron/ion two-stream instability (Buneman, 1959). A representative dispersion plot is shown in Figure 3.5. Although this figure is computed from the linear Vlasov dispersion equation, it is very similar to the dispersion curve obtained from cold plasma theory, illustrating the strong dispersion and narrow range of wavenumber growth characteristic of that limit. At maximum growth of the Buneman instability, $k_z v_{oe} \simeq \omega_e$.

Figure 3.6 illustrates the maximum growth rate of these electron/ion instabilities as a function of the electron/ion drift speed for $T_e = 10T_i$

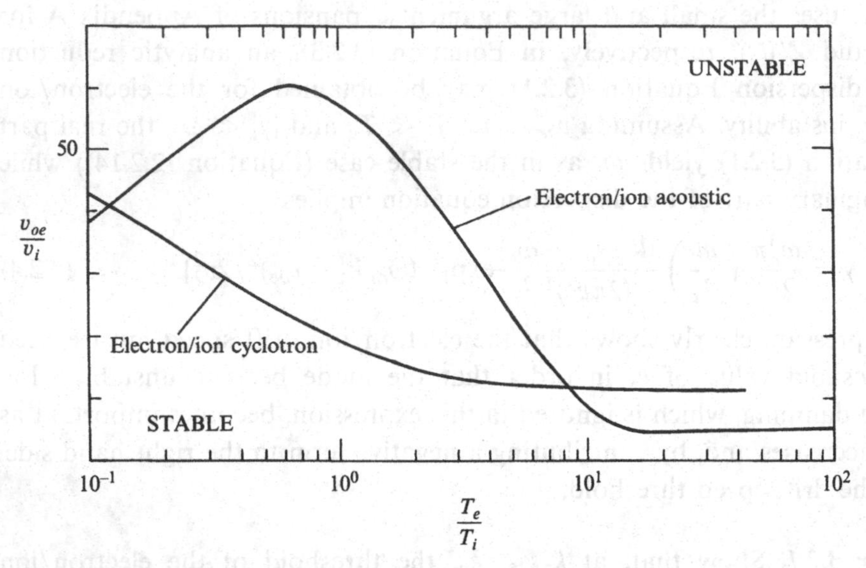

Fig. 3.4 The threshold electron/ion drift speeds for the electron/ion acoustic instability and the electron/ion cyclotron instability as functions of T_e/T_i. Here the cyclotron instability propagates in a magnetized plasma with $\omega_e = 10|\Omega_e|$; this mode is discussed in Subsection 3.3.1.

and $T_e = T_i$. This figure shows how the electron/ion acoustic instability is enhanced at relatively small v_{oe} by $T_e \gg T_i$, but also demonstrates how this mode goes over to the electron/ion two-stream instability in the $v_e \ll v_{oe}$ limit. Using the asymptotic expansion of the plasma dispersion function (Equation (A.8) of Appendix A) for both the electron and ion susceptibilities of Equation (3.2.3), the dispersion equation (3.2.1) becomes

$$1 - \frac{\omega_i^2}{\omega^2} - \frac{\omega_e^2}{(\omega - k_z v_{oe})^2} = 0. \tag{3.2.5}$$

From this equation it follows that the maximum growth rate of the electron/ion two-stream instability is

$$\gamma_m \simeq \frac{3^{1/2}}{2^{4/3}} \left(\frac{m_i}{m_e}\right)^{1/6} \omega_i. \tag{3.2.6}$$

Because $(m_i/m_e)^{1/6}$ is such a weak function of the mass ratio, $\gamma_m \sim \omega_i$ is a good rule of thumb for this instability.

Problem 3.2.2. Derive Equation (3.2.6) from (3.2.5). Hint: the crucial steps here are, first, to note $k_z v_{oe} \simeq \omega_e$ at γ_m in Figure 3.5 and use this relation in (3.2.5), and second, to observe $\omega_r \ll \omega_e$ from the same figure and use this to expand the resulting term.

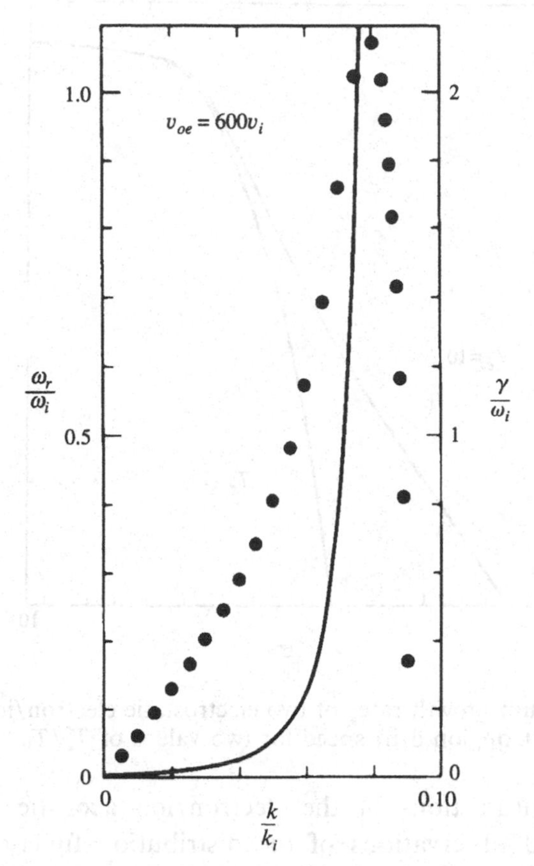

Fig. 3.5 The real frequency (solid line) and growth rate (dotted line) of the Buneman instability as functions of wavenumber. Here $T_e = 10T_i$, $v_{oi} = 0$ and $v_{oe} = 600v_i = 4.43v_e$.

Applications. We define ion-acoustic-like fluctuations as those observed to be essentially electrostatic (no detectable fluctuating magnetic field) and in the vicinity of the (Doppler shifted) ion plasma frequency. Enhanced fluctuations that satisfy these conditions have been observed in many different space plasma contexts, including the solar wind (Gurnett, 1991; Marsch, 1991, and references therein), at and near the Earth's bow shock (Gurnett, 1985, and references therein), and the Earth's magnetotail. Although ion-acoustic-like fluctuations are frequently observed, there is a similarly frequent problem with their interpretation in terms of the electron/ion acoustic instability: observations typically indicate $T_e \sim T_i$ and $v_{oe} \ll v_e$, conditions inappropriate for the excitation of such waves.

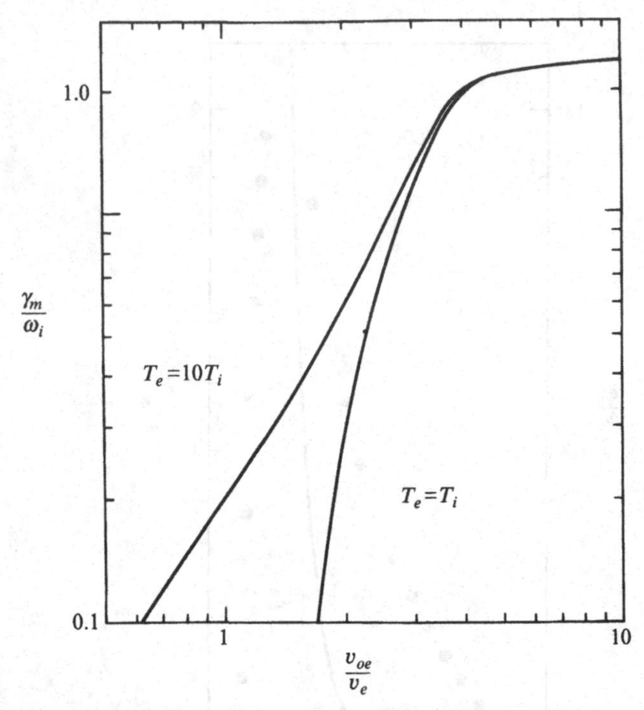

Fig. 3.6 The maximum growth rates of two electrostatic electron/ion instabilities as functions of the electron/ion drift speed for two values of T_e/T_i.

Theoretical identifications of the electron/ion acoustic instability supported by detailed observations of the distribution functions seem to be confined thus far to individual cases in the Earth's bow shock (Thomsen *et al.*, 1983) and in the solar wind (Dum *et al.*, 1980) during which the exceptional condition $T_i \ll T_e$ obtains. Comparative studies of instability thresholds in the solar wind (Gary, 1978; Lemons *et al.*, 1979) indicate that both the whistler heat flux instability and the ion/ion acoustic instability often have lower thresholds than the electron/ion acoustic instability (see Subsection 3.2.4). Thus the latter mode may not be found very often in the solar wind or other space plasmas.

3.2.2 High frequency electron/electron instabilities

In this subsection we consider the plasma to consist of three components: ions (subscript *i*), an electron beam (*b*) and an electron core (*c*). The ions are at rest in the center-of-mass frame ($v_{oi} = 0$) and the electrons bear zero net current: $n_c\mathbf{v}_{oc} + n_b\mathbf{v}_{ob} = 0$.

In the presence of a relatively weak, relatively slow electron beam, there are two high frequency ($\omega_i \ll \omega_r$) normal modes of the plasma: an electron plasma (or Langmuir) wave with real frequency satisfying Equation (2.2.13) and an electron beam wave:

$$\omega_r \simeq k_z v_{ob}. \tag{3.2.7}$$

As v_o is increased, one of these two waves becomes unstable. The criterion for distinguishing which was given by O'Neil and Malmberg (1968) (modified here to correspond to a Maxwellian beam):

$$\left(\frac{v_{ob}}{v_b}\right)^3 \left(\frac{n_b}{n_e}\right) < 1 \tag{3.2.8a}$$

for the Langmuir beam instability and

$$\left(\frac{v_{ob}}{v_b}\right)^3 \left(\frac{n_b}{n_e}\right) \gtrsim 1 \tag{3.2.8b}$$

for the electron/electron beam instability.

In the limit of weak beam density and low beam speeds it is the Langmuir wave that first goes unstable, as is illustrated by the weak growth at $k/k_i \simeq 0.25$ in the left hand panel of Figure 3.7. Note that although the beam mode crosses the Langmuir wave in the ω_r vs k diagram, a condition that often leads to wave growth in the fluid approximation, the instability does not arise at this crossing. Rather, at these relatively low drift speeds both modes are kinetic, the damping of the beam mode is much greater than that of the Langmuir wave and the frequency degeneracy that leads to fluid instabilities is absent. In other words, near threshold the modes are both kinetic and the condition for instability is that the slope of the distribution function must be positive at the phase speed of the wave, a more stringent condition than that of the simple crossing criterion of fluid theory. For the parameters of Figure 3.7, instability threshold corresponds to $v_{ob} = 4.67 v_e$, considerably larger than that necessary to yield a crossing of the two modes.

As the drift speed is increased, the mode topology changes and the waves do interact. As illustrated in the right hand panel of Figure 3.7, the beam mode at small k takes on Langmuir dispersion at large k and vice versa. The instability makes a transition to the the lower frequency mode whereas the upper frequency branch becomes strongly damped. As the O'Neil-Malmberg criterion indicates, the parameters here correspond to the transition between the Langmuir and electron/electron beam instabilities. Increasing the drift speed further increases the Doppler shift of the beam mode and reduces the wavenumber at maximum growth rate. However, for this very tenuous beam

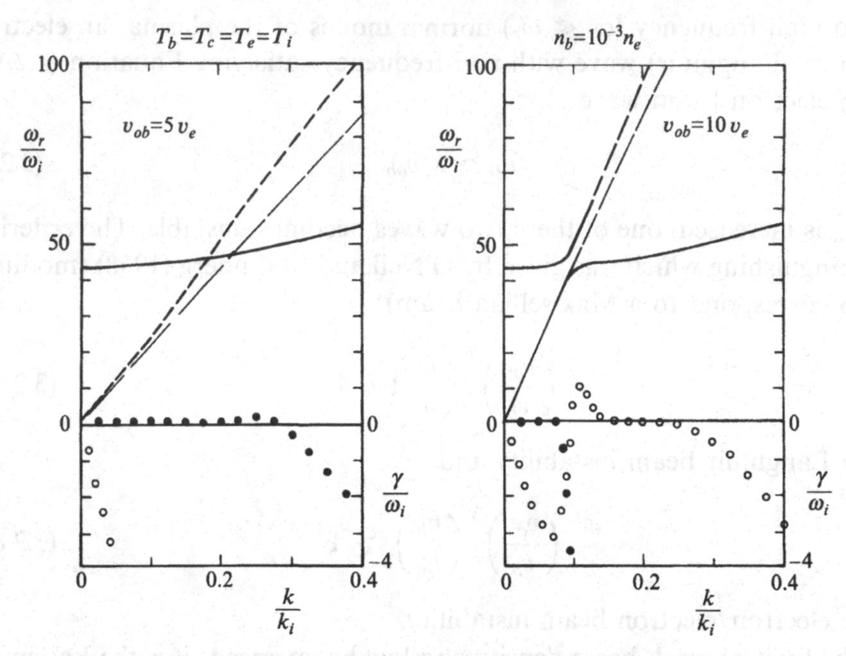

Fig. 3.7 The real frequencies (solid/short dashed lines) and growth rates (dotted lines) of the Langmuir and electron beam modes as functions of wavenumber for two different values of the beam drift speed. Here, and in Figures 3.8 through 3.10, the plasma consists of three components: the ions (i), an electron core (c) and an electron beam (b). A short dashed line indicates that the mode is heavily damped ($\gamma < -|\omega_r|/2\pi$). The long dashed lines represent $\omega_r = k_z v_{ob}$. The lines of solid dots represent γ of the mode that is Langmuir-like at small wavenumber; the lines of open dots represent the damping and growth of the mode that is beam-like at small k. Here $T_b = T_c = T_e = T_i$ and $n_b = 0.001 n_e$.

case, the instability remains essentially confined to $\omega_r \simeq \omega_e$, and it would be experimentally difficult to resolve the two different modes.

The electron/electron beam instability appears more clearly at larger beam densities, as illustrated in Figure 3.8. For the 10% beam used here, it is the lower frequency branch that becomes unstable at threshold and remains the unstable mode as the drift speed increases. Because this electron/electron beam instability satisfies $\omega_r \simeq k_z v_{ob}$, it grows via a Landau resonance ($|\zeta_b| \lesssim 1$) with the beam. In addition, as Figure 3.8 illustrates, for a sufficiently dense beam, fluctuations may grow across the broad range of frequencies $\omega_i < \omega_r \lesssim \omega_e$ and, in particular, ω_r can be much less than ω_e at maximum growth rate. This issue has been studied by Fuselier *et al.* (1985) and Gary (1985).

The threshold of the electron/electron beam instability is a function of both the beam density and beam temperature, as is illustrated in Figure 3.9.

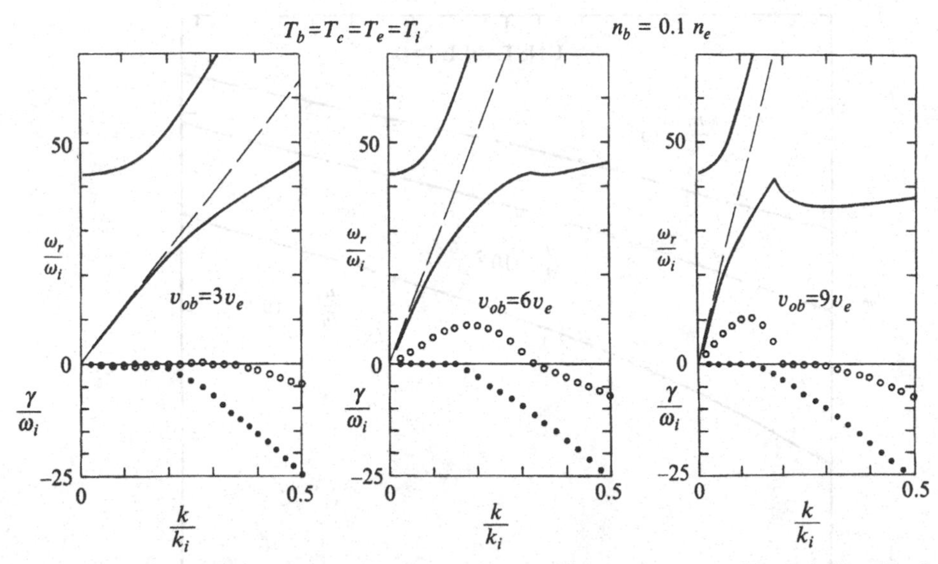

Fig. 3.8 The real frequencies (solid lines) and growth rates (dotted lines) of the Langmuir-like and electron beam modes as functions of wavenumber for three different values of the beam drift speed. The long dashed lines represent $\omega_r = k_z v_{ob}$. The lines of solid dots represent γ of the mode that is Langmuir-like at small wavenumber; the lines of open dots represent the damping and growth of the mode that is beam-like at small k. Here $T_b = T_c = T_e = T_i$ and $n_b = 0.10 n_e$. (From Fig. 1 of Gary, 1985)

Increasing the parameter $(n_b/n_e)(T_c/T_b)$ acts to increase the slope of the beam and thereby lowers the threshold.

As v_o increases, γ_m generally increases and the wavenumber at maximum growth decreases. The maximum growth rate as a function of the drift speed is shown in Figure 3.10 for both the electron/electron beam instability and the electron/ion acoustic instability. As the relative drift between the two components exceeds their thermal speeds, the components usually become nonresonant ($1 \ll |\zeta_j|$) and the cold plasma approximation becomes valid. In this nonresonant limit, the electron/electron beam instability is called the electron/electron two-stream instability. Because the dispersion equation in the cold plasma limit is similar to that of the electron/ion two-stream instability, the procedure for solving it is almost the same so that, at $n_b \ll n_c$ for the electron/electron two-stream instability

$$\gamma_m = \frac{3^{1/2}}{2}\left(\frac{n_b}{2n_c}\right)^{1/3}\omega_c, \tag{3.2.9}$$

which is in approximate agreement with the large drift speed results of Figure 3.10.

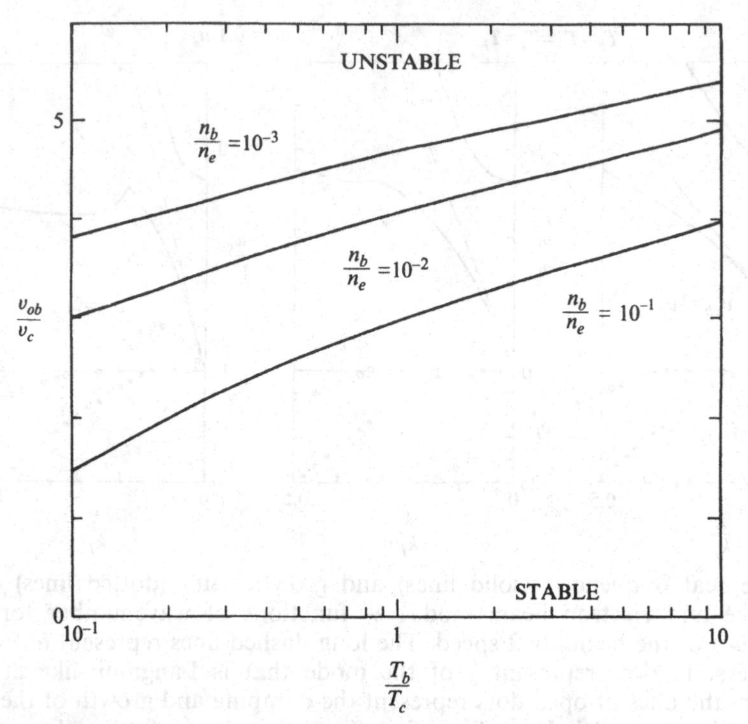

Fig. 3.9 The threshold beam/ion drift speed for the electron/electron beam instability as a function of T_b/T_c for three different values of n_b/n_e.

Problem 3.2.3. Show that, by translating into the reference frame of the beam, one may use approximations completely analogous to those of Problem 3.2.2 and that Equation (3.2.9) is the resulting expression for the maximum growth rate of the electron/electron two-stream instability.

Under the condition of zero current, both the beam and core electron components have a drift speed relative to the ions. Thus, not only may the beam/core relative drift give rise to an electron/electron instability, but the core/ion drift may excite an electron/ion instability. The relative thresholds of the two instabilities have been examined by Gary (1985); the results of that study show that the electron/ion acoustic instability has the lower threshold only if the core electrons are considerably hotter than the ions or, if $T_e \simeq T_i$, when $n_c \sim n_b$. Because neither of these conditions is frequently found in space plasmas, we once again conclude that the electron/ion acoustic instability is not an important contributor to space plasma instabilities.

Applications. Large amplitude electric field fluctuations near the electron plasma frequency are frequently observed in space plasmas. Observations

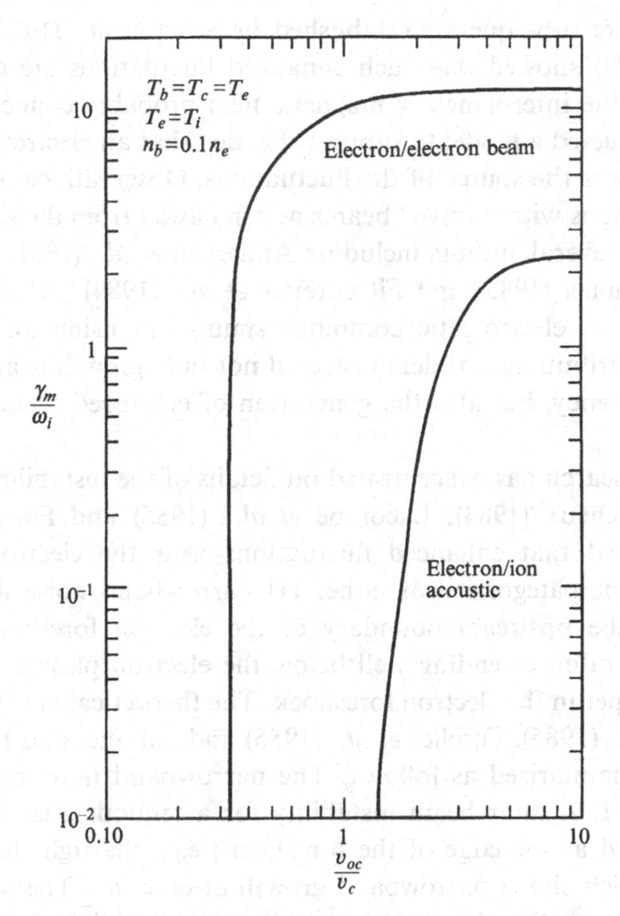

Fig. 3.10 The maximum growth rates of the electron/electron beam and electron/ion acoustic instabilities as functions of the electron core/ion relative drift speed. Here $n_b = 0.10n_e$ and $T_b = T_c = T_e = T_i$. (From Fig. 3 of Gary, 1985)

and interpretations of ω_e fluctuations in the solar wind have been reviewed by Gurnett (1991). Relatively cold electron beams that are likely to give rise to electron/electron instabilities have been observed in the low latitude boundary layer near the magnetopause (Gosling *et al.*, 1990) and in the plasma sheet boundary layer (Onsager *et al.*, 1990). Here we discuss the plasma regime upstream of the Earth's bow shock, where the identification of such fluctuations with electron beams is now well established.

Electrons accelerated near the quasiperpendicular part of the terrestrial bow shock escape back upstream into the solar wind and are observed in a region termed the electron foreshock. Enhanced fluctuations near the plasma frequency were first observed in the electron foreshock by Fredricks *et al.* (1968), and correlations of these fluctuations with enhanced fluxes of energetic electrons were subsequently established by Scarf *et al.* (1971). Filbert and Kellogg (1979) showed that such enhanced fluctuations are correlated with times when the interplanetary magnetic field probably connects to the shock, and constructed a model to support the idea that an electron/electron beam instability was the source of the fluctuations. Observations of electron distribution functions with resolved beams moving away from the shock have been reported by several authors including Anderson *et al.* (1981), Feldman *et al.* (1983), Klimas (1983) and Fitzenreiter *et al.* (1984). Klimas (1983) also carried out an electrostatic computer simulation using an observed electron beam distribution that demonstrated not only growth near the electron plasma frequency, but also the generation of enhanced fluctuations at harmonics of ω_e.

More recent research has concentrated on details of the instability process. Etcheto and Faucheux (1984), Lacombe *et al.* (1985) and Fuselier *et al.* (1985) all observed that enhanced fluctuations near the electron plasma frequency could be categorized as either (1) narrowband noise just above ω_e observed at the upstream boundary of the electron foreshock, or (2) broadband noise often extending well below the electron plasma frequency and observed deeper in the electron foreshock. The theoretical interpretations of Lacombe *et al.* (1985), Fuselier *et al.* (1985) and Onsager and Holzworth (1990) may be summarized as follows: The narrowband noise can be well explained by the Langmuir beam instability for a tenuous, fast beam that would be expected at the edge of the foreshock; e.g., the right hand panel of Figure 3.7, which shows narrowband growth at $\omega_r \simeq \omega_e$. The broadband noise corresponds to the electron/electron beam instability for a slower, more dense beam that would be expected deeper in the foreshock; Figure 3.8 shows the corresponding growth rates are nonzero over a much broader range of frequencies including $\omega_r \ll \omega_e$.

3.2.3 The electron/electron acoustic instability

In this subsection we again consider the configuration of two drifting electron components and a single ion component such that the net current is zero. However, because we are concerned with parameters appropriate to the electron acoustic wave of Subsection 2.2.3, we assume that one electron component is hot (subscript h) and the other cool (c).

If $T_c \ll T_h$ and there are no relative drifts among the components, three kinds of lightly damped normal modes may propagate: Langmuir waves at $\omega_e < \omega_r$ (Subsection 2.2.1), ion acoustic waves at $\omega_r < \omega_i$ (Subsection 2.2.2), and electron acoustic waves at $\omega_r \sim \omega_c$ (Subsection 2.2.3). As v_o,

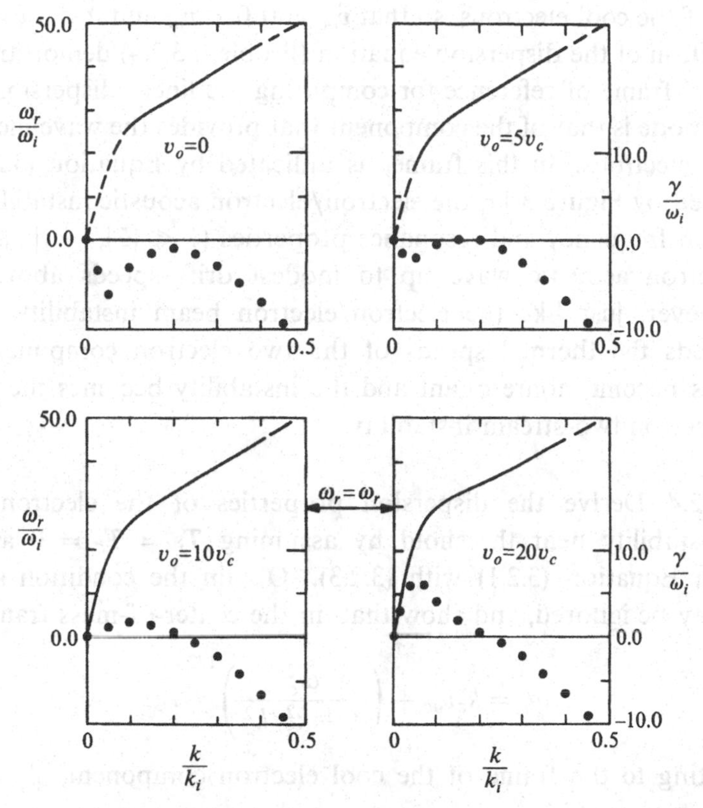

Fig. 3.11 The real frequencies (solid/dashed lines) and growth rates (dotted lines) of the electron acoustic wave and electron/electron acoustic instability as functions of wavenumber for four different values of the hot/cool relative drift speed. Here, and in Figures 3.12 and 3.13, the plasma consists of three components: the ions (i), the hot electrons (h) and the cool electrons (c). The results here are computed in the frame of the cool electrons ($v_{oc} = 0$ and $v_{oh} = v_o$). The dashed lines indicate that the mode is heavily damped ($\gamma < -|\omega_r|/2\pi$). Here, and in Figures 3.12 and 3.13, $T_c = T_i$ and $T_h = 100T_c$. Here $n_h = n_c = n_e/2$. (From Fig. 3 of Gary, 1987).

the hot/cool relative drift speed, is increased subject to the zero current condition, two modes eventually become unstable: the electron/ion acoustic instability that propagates in the direction of the cool electron drift velocity, and the electron/electron acoustic instability that propagates with $\mathbf{k}_m$ parallel to $\mathbf{v}_{oh}$. The dispersion properties of the latter mode are illustrated in Figure 3.11 for four different values of v_o. As for other instabilities examined in this chapter, an increase in the relative drift speed reduces the damping rate until the instability threshold at $\gamma = 0$ is reached; further increases in v_o lead to instability growth.

It is important to note that the results of Figure 3.11 are computed in the rest frame of the cool electrons, so that $v_{oc} = 0, 0 < v_{oi}$ and $0 < v_{oh} = v_o$. Analytic reduction of the dispersion equation (Problem 3.2.4) demonstrates that the preferred frame of reference for computing the linear dispersion properties of this mode is that of the component that provides the wave inertia, that is, the cool electrons. In this frame, as indicated by Equation (3.2.10) and demonstrated by Figure 3.11, the electron/electron acoustic instability maintains the real frequency and resonance properties ($1 \ll |\zeta_i|, 1 < |\zeta_c|, |\zeta_h| < 1$) of the electron acoustic wave up to modest drift speeds above thresholds. However, just like the electron/electron beam instability, when v_o much exceeds the thermal speeds of the two electron components, both components become nonresonant and the instability becomes the fluid-like electron/electron two-stream instability.

Problem 3.2.4. Derive the dispersion properties of the electron/electron acoustic instability near threshold by assuming $T_i = T_c = 0$ and using $|\zeta_h| \ll 1$ in Equation (3.2.1) with (3.2.3). Obtain the condition such that $K_i(\mathbf{k}, \omega)$ may be ignored, and show that, in the center-of-mass frame,

$$\omega_r = k_z v_{oc} + \left(\frac{\omega_c^2}{1 + k_h^2/k^2} \right)^{1/2}. \tag{3.2.10}$$

By translating to the frame of the cool electron component, show that, if $\gamma \ll \omega_r$,

$$\gamma \simeq \frac{\omega_c}{2} \frac{k_h^2}{k^2} \left(\frac{\pi}{2} \right)^{1/2} \left(\frac{v_o - \omega_r/k_z}{v_h} \right) \exp\left[-\frac{(v_o - \omega_r/k_z)^2}{2v_h^2} \right]. \tag{3.2.11}$$

Because $k \simeq 5k_c$ at threshold, show that $v_c \ll v_o$ must obtain in order that this mode become unstable.

Figure 3.12 shows the core density dependence of the threshold hot/cool relative drift speed for both the electron/electron and electron/ion acoustic instabilities. At $T_c \simeq T_i$ and under the zero current condition, the electron/ion acoustic instability has a much lower threshold than the electron/electron acoustic instability at $n_c \lesssim 0.70 n_e$. An increase in T_c/T_i enhances the electron/ion acoustic instability and increases the parameter regime in which it has the lower threshold; a decrease in T_c/T_i has the opposite effect. However, if the configuration is changed by removing the zero current condition and setting $v_{oc} = v_{oi} = 0$, then the electron/ion acoustic instability is suppressed but the electron/electron acoustic unstable mode remains essentially unchanged. Thus for a relatively large range of n_c/n_e values, the observation of enhanced electron acoustic noise at frequencies

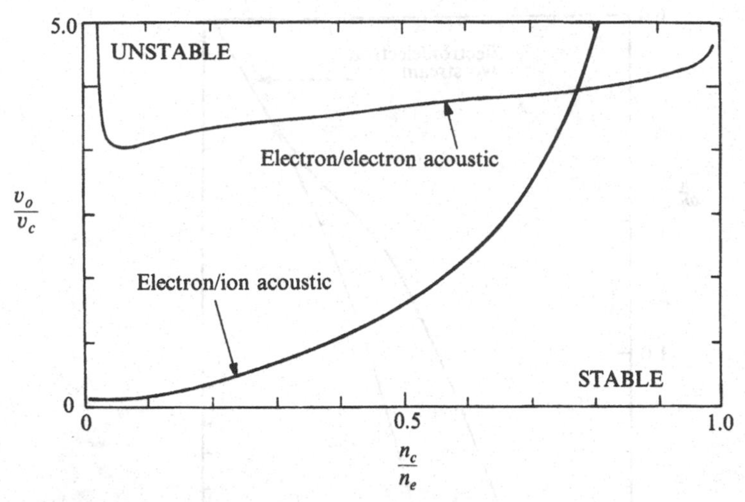

Fig. 3.12 The threshold hot/cool relative drift speeds for the electron/electron acoustic and electron/ion acoustic instabilities as functions of the cool electron density. (From Fig. 6 of Gary, 1987).

well above ω_i indicates the existence of a strong field-aligned current; this is indeed the case for the observations of cusp hiss discussed below.

Figure 3.13 illustrates the maximum growth rate of instabilities in the zero current configuration as a function of the hot/cool relative drift speed. Not only does the electron/ion acoustic instability have the lower threshold, but for these parameters it maintains a larger growth rate than the electron/electron acoustic instability until $v_h < v_{oh}$ where the two modes merge into the nonresonant electron/electron two-stream instability. Again, if $v_{oc} = v_{oi} = 0$, then the electron/ion acoustic instability is suppressed and the electron/electron acoustic instability is the mode most likely to be observed under the condition of nonzero current.

Applications. The electron/electron acoustic instability has been studied at the Earth's bow shock (Thomsen *et al.*, 1983) and upstream of the bow shock (Marsch, 1985). The most detailed and convincing theory of this instability concerns its application to the phenomenon termed "cusp hiss."

Burch *et al.* (1983) used measurements from the Dynamics Explorer 1 satellite to demonstrate that electron beams moving upward with energies $20eV \lesssim E \lesssim 200eV$ are a common feature of the region equatorward of and within the polar cusp. Lin *et al.* (1984) then demonstrated a correlation between these beams and "cusp hiss," intense electric fluctuations at several kilohertz.

Lin *et al.* also solved the electrostatic Vlasov dispersion equation for a

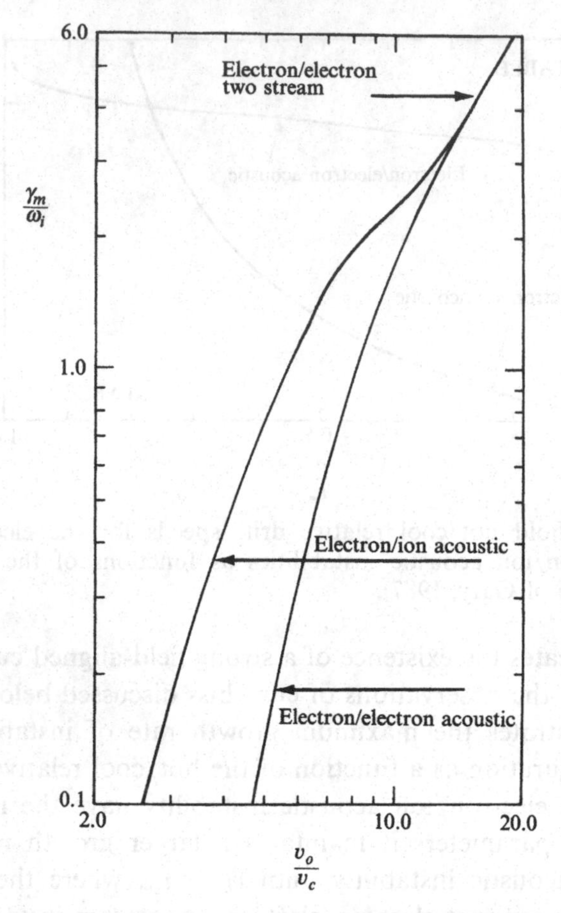

Fig. 3.13 The maximum growth rates of the electron/electron acoustic, electron/ion acoustic and electron/electron two-stream instabilities as functions of the relative drift speed between the hot and cool electron components. Here $n_h = n_c = n_e/2$ (From Fig. 8 of Gary, 1987).

plasma with three electron components consisting of a cold background, a hot background and a hot beam with drift speed parallel to $\mathbf{B}_o$. They demonstrated that an electron beam could drive a mode unstable at the observed frequencies and with maximum growth rate at propagation parallel to $\mathbf{B}_o$. Lin *et al.* identified this unstable mode as a whistler. However, Tokar and Gary (1984) used the more general electromagnetic Vlasov dispersion equation to show that the whistler mode is stable for the model parameters of Lin *et al.* (1984), and that the unstable mode is the electron/electron acoustic instability.

Lin *et al.* (1985) carried out computer simulations of this growing mode using the three-component electron model. Their results show that, although the instability saturation mechanism is a sensitive function of the model parameters, in all cases reported the cold background electrons were heated much more strongly than either of the two hot components. Because the existence of the electron/electron acoustic instability critically depends on the existence of a relatively dense, very cold component, it seems that, for the simulations, the basic saturation mechanism of this mode is heating of the cold component. Although the simulation code used in Lin *et al.* (1985) is fully electromagnetic, no excitation of electromagnetic fluctuations was observed. Thus a major unanswered question here is the source of the frequently observed electromagnetic component of cusp hiss.

3.2.4 Ion/ion acoustic instability

In this subsection we consider the configuration of an ion beam (subscript b), an ion core (c) and electrons (e). There is a beam/core relative drift $\mathbf{v}_o$, but the electron drift speed is chosen to maintain zero current. In this configuration, there are two modes that may become unstable as v_o is increased: the electron/ion acoustic instability driven by the ion core/electron relative drift and propagating in the direction of $\mathbf{v}_{oc}$ in the electron frame, and the ion/ion acoustic instability driven by the beam/core relative drift and propagating parallel to $\mathbf{v}_{ob}$ in the center-of-mass frame.

Figure 3.14 illustrates dispersion properties of ion-acoustic-like modes that propagate in the direction of the beam velocity as the beam/core relative drift speed is increased from zero. At $v_o = 0$, the only lightly damped low frequency mode in the plasma is the ion acoustic wave of Subsection 2.2.2. As the beam drift speed increases, however, the phase speed of the wave begins to increase as does the magnitude of the damping rate. At the same time, however, one of the heavily damped acoustic-like modes mentioned in Subsection 2.2.2 experiences a reduction in its damping rate, so that at sufficiently large v_o, the mode becomes unstable. This instability is sometimes called the ion/ion streaming instability at $n_b = n_c$. But because ω_r is typically proportional to the instability wavenumber, we term this mode the ion/ion acoustic instability (Fried and Wong, 1966; Imre and Ozizmir, 1974; Fried and Wong, 1974; Gary and Omidi, 1987).

As for the ion acoustic wave, the electrons are Landau resonant ($|\zeta_e| \ll 1$), and the core ions relatively nonresonant ($|\zeta_c| > 1$) with the ion/ion acoustic instability. The ion beam, however, is typically Landau resonant near the instability threshold ($|\zeta_b| \lesssim 1$) and becomes nonresonant only at relatively high beam/core relative drift speeds ($v_b \ll v_{ob}$). This beam resonance prevents either a series or an asymptotic expansion of the $Z'(\zeta_b)$ factor in

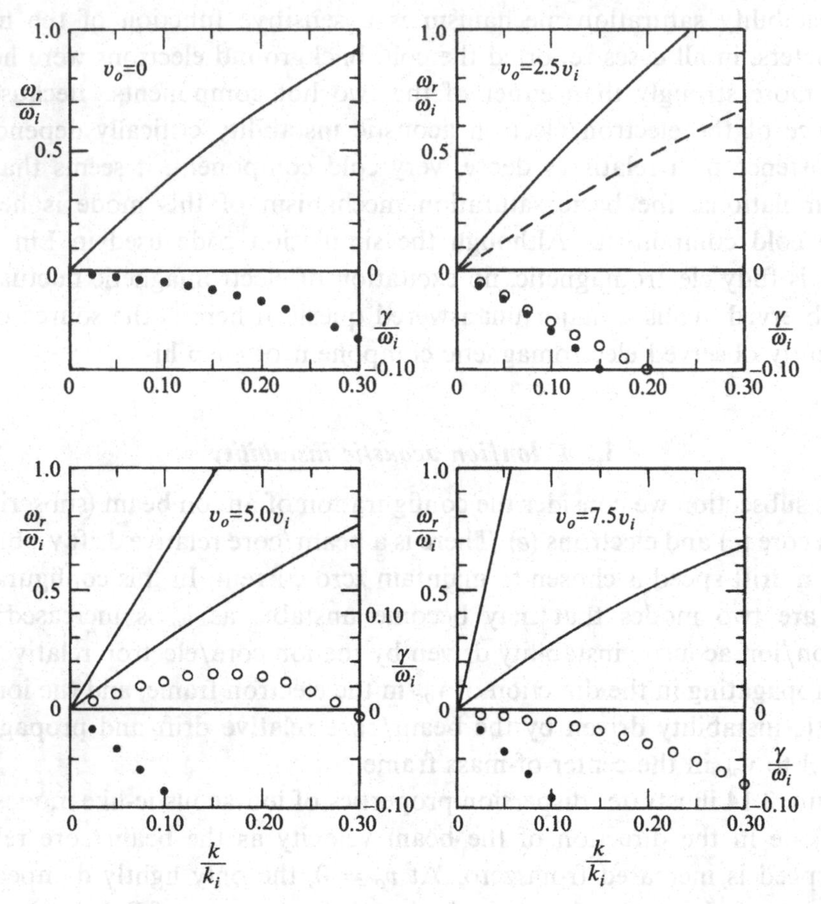

Fig. 3.14 The real frequencies (solid/dashed lines) and damping/growth rates (dotted lines) of the ion acoustic modes and ion/ion acoustic instability as functions of wavenumber for four different values of the ion beam/core relative drift. Here and in Figures 3.15 through 3.17 the plasma consists of three components: the electrons (*e*), a more dense ion core (*c*) and a less dense ion beam (*b*). The dashed lines indicate that the mode is heavily damped ($\gamma < -|\omega_r|/2\pi$). At zero drift speed, the only lightly damped mode is the ion acoustic wave (see Figure 2.3). As v_o increases, this mode evolves into the root of greater phase speed with damping represented by the solid dots. The mode of lower phase speed has γ represented by the open dots and corresponds to the ion/ion acoustic instability at $v_o = 5v_i$. Here $n_b = 0.10n_e$, $T_b = T_c = T_i$ and $T_e = 10T_i$. (From Fig. 1 of Gary and Omidi, 1987).

the beam susceptibility and, because the beam contribution to the dispersion equation is often non-negligible, has prevented us from obtaining an analytic expression for ω_r that agrees with the computed results for any broad parameter range. Gary and Omidi (1987) empirically find at $T_b = T_c = T_i$

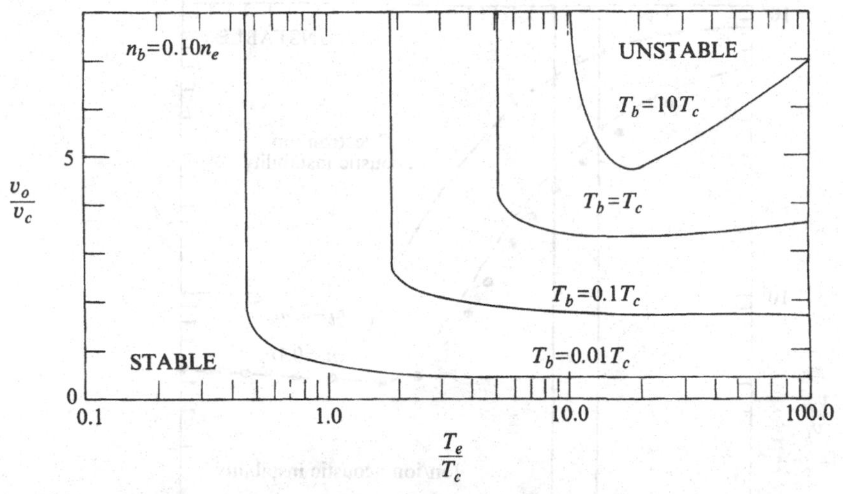

Fig. 3.15 The threshold beam/core relative drift speed for the ion/ion acoustic instability as a function of T_e/T_c for four different values of the beam/core temperature ratio. Here $n_b = 0.10n_e$. (From Fig. 6 of Gary and Omidi, 1987).

and $m_b = m_c = m_i$

$$\frac{\omega_r}{k} \simeq c_s \left(\frac{n_c - n_b}{n_e} \right). \tag{3.2.12}$$

But even with $n_b \ll n_c$, ω_r/k falls well below both c_s and v_i as T_b/T_c becomes much less than unity.

Figure 3.15 shows the threshold drift speed of this instability as a function of T_e/T_c for four different values of T_b/T_c. Just as for the electron/electron beam instability (Figure 3.9), a reduction of the beam temperature significantly reduces the threshold drift speed of the ion/ion acoustic instability. Note also that, because $\omega_r/k_z < v_c$ at $T_b \ll T_c$, the threshold beam/core drift speed may also lie below the core thermal speed when the beam temperature is sufficiently small.

Another important property of the ion/ion acoustic instability illustrated in Figure 3.15 is the T_e/T_c cutoff; below a certain value of this temperature ratio for a fixed n_b/n_i and T_b/T_c the instability is stabilized for all drift speeds, no matter how large v_o becomes. Of course, increasing the beam density or decreasing T_b/T_c reduces the threshold drift speed and the T_e/T_c cutoff (Fried and Wong, 1966; Gary and Omidi, 1987). Similarly, an increasing beam density or increasing T_e/T_i enhances the maximum growth rate (Abas and Gary, 1971).

If T_e/T_c is greater than the cutoff value, the ion/ion acoustic instability typically has a much lower threshold than its electron/ion counterpart, and

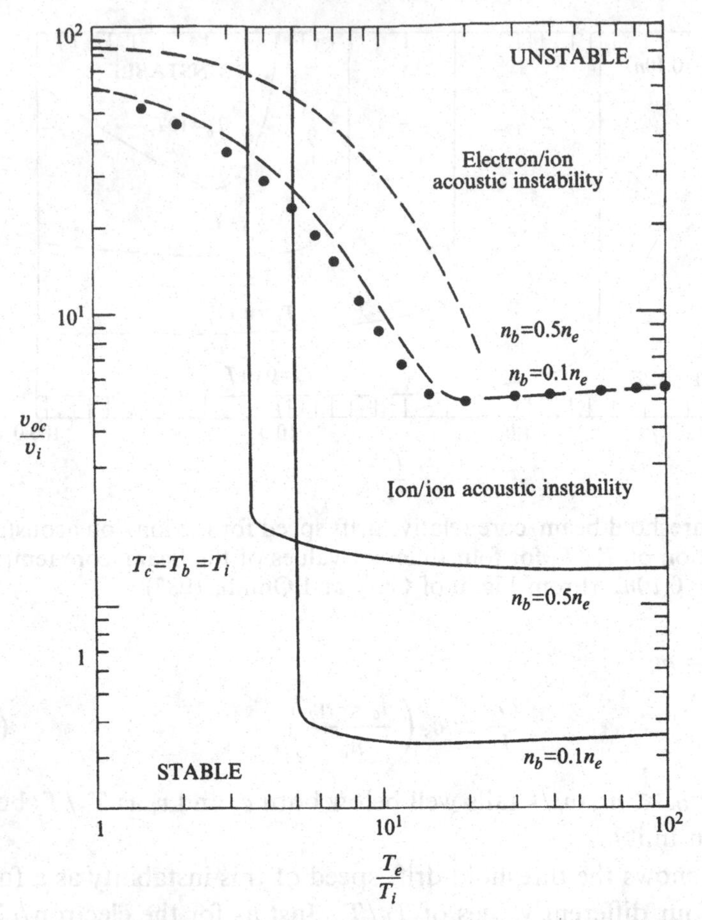

Fig. 3.16 The threshold core drift speeds for ion acoustic instabilities as functions of T_e/T_i. The solid lines represent the ion/ion acoustic instability at two different values of the ion beam density. The dashed lines represent the electron/ion acoustic instability in the two-ion component configuration for two different values of the ion beam density. The dotted line represents the electron/ion acoustic instability at zero beam density and nonzero current. Here $n_b = 0.10 n_e$ and $T_b = T_c = T_i$. (From Fig. 5 of Gary and Omidi, 1987).

may be the more important mode at relatively low drift speeds. This is illustrated by Figure 3.16, which compares the thresholds of the ion/ion and electron/ion acoustic instabilities at $T_b = T_c$. We offer the following explanation for this result. The electron/ion acoustic instability is driven by the electron/core relative drift speed that must satisfy $c_s < v_{oc}$ for fluctuation growth. In contrast, the ion/ion acoustic instability is driven by the beam/core relative drift speed, with instability condition $c_s \lesssim v_o$. For the latter instability, if $n_b \ll n_c, v_{oc} \ll v_{ob}$ under the zero current condition and

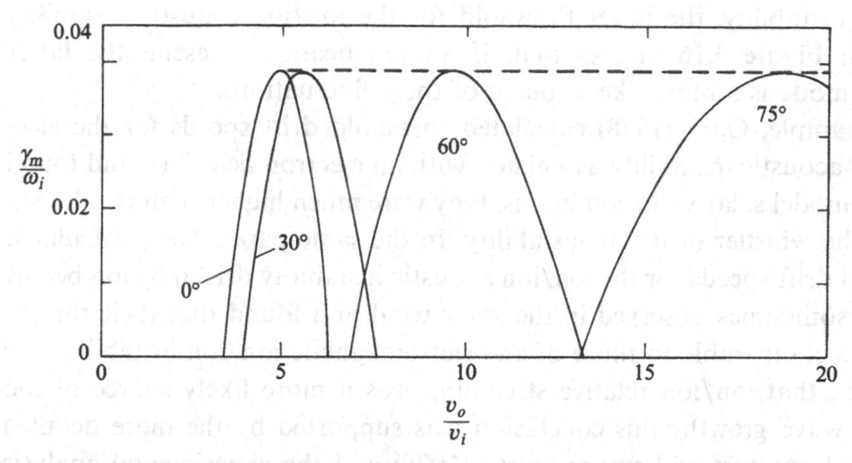

Fig. 3.17 The maximum growth rate of the ion/ion acoustic instability as a function of the beam/core relative drift speed at four different angles of propagation. The dashed line indicates maximum growth rate for a given v_o independent of angle. Here $n_b = 0.10 n_e$, $T_b = T_c = T_i$ and $T_e = 10 T_i$. (From Fig. 2 of Gary and Omidi, 1987).

the core drift speed requirement is much less stringent than the condition for the electron/ion instability. Furthermore, if $T_b < T_c$, v_o need not exceed c_s in order that the ion/ion acoustic mode be unstable, reducing the threshold condition on v_{oc} even further.

As v_o is increased above threshold, the growth rate of the ion/ion acoustic instability reaches a maximum and then, at a sufficiently large value of the drift speed, the mode returns to a stable condition at $\mathbf{k} \times \mathbf{v}_o = 0$ (Fried and Wong, 1966). This stabilization at large v_o is quite different from the response of the electron/ion acoustic instability, which has a maximum growth rate increasing monotonically with v_o (e.g. Figure 3.10).

What happens here is that as v_o increases above the value necessary to stabilize this mode at $\mathbf{k}$ parallel to $\mathbf{v}_{ob}$, the instability persists, but only at successively steeper angles of propagation. This peculiar feature is illustrated in Figure 3.17, which also shows that oblique propagation does not further enhance γ_m over its value at parallel propagation. Thus the ion/ion acoustic instability never enters the nonresonant regime at high v_o values and we expect that the electron/ion acoustic instability driven by the ion core/electron relative drift dominates the ion/ion instability at sufficiently high drift speed.

Applications. As we mentioned earlier, ion-acoustic-like fluctuations have been observed in many different space plasma regimes. Although it has been difficult to establish that these fluctuations correspond to the electron/ion acoustic instability, the lower threshold for the ion/ion acoustic instability shown in Figure 3.16 implies that, if an ion beam is present, the latter growing mode is a more likely source of these fluctuations.

For example, Gary (1978) calculated threshold drift speeds for the electron/ion acoustic instability associated with an electron heat flux, and found that, for model solar wind conditions, they were much higher than the thresholds of the whistler heat flux instability. In the same paper, Gary calculated threshold drift speeds for the ion/ion acoustic instability driven by ion beams that are sometimes observed in the solar wind and found that their thresholds were comparable to those of an electromagnetic ion/ion instability. He concluded that ion/ion relative streaming was a more likely source of ion acoustic wave growth; this conclusion was supported by the more detailed theoretical analysis of Lemons *et al.* (1979) and the experimental analysis of Kurth *et al.* (1979).

The evidence that the ion/ion acoustic instability operates at and near the Earth's bow shock is more substantial. Fuselier and Gurnett (1984) studied ion-acoustic-like noise observed upstream of the Earth's bow shock. Through measurements of both wavelengths and frequencies, Fuselier and Gurnett were able to show that the fluctuations approximately satisfied the ion acoustic dispersion relation. They were also able to show that the fluctuations propagated in the sunward direction, opposite the solar wind velocity, and obliquely, rather than parallel, to the background magnetic field; both of these properties are consistent with the ion/ion acoustic instability. Greenstadt and Mellott (1987) showed that the observed rise of ion-acoustic-like noise immediately upstream of the bow shock scales as the expected gyroradius of reflected ions at the shock, lending further support to the idea that the ion/ion acoustic instability is operating there.

Another application of the ion/ion acoustic instability has been to the phenomenon called broadband electrostatic noise discovered by Scarf *et al.* (1974) in the magnetotail. Broadband electrostatic noise was first studied in detail by Gurnett *et al.* (1976) who showed that these enhanced fluctuations often are found in the plasma sheet boundary layer, the transition region between the low β lobe plasma and the high β plasma sheet. Gurnett *et al.* (1976) also established that the fluctuating electric field of broadband electrostatic noise is oriented at a relatively steep angle to the ambient magnetic field, and that the phenomena often occurs in regions that have highly anisotropic fluxes of streaming protons.

Grabbe and Eastman (1984) showed that broadband electrostatic noise often is correlated with a relatively dense, cold ion component that streams parallel to the magnetic field in the boundary layer. It was soon established that the resulting instability has its maximum growth rate at a relatively steep angle to the ambient magnetic field (Grabbe, 1985; Omidi, 1985), and that it is therefore the ion/ion acoustic instability (Akimoto and Omidi, 1986).

To account for broadband electrostatic noise observed at frequencies above 1 kilohertz, Grabbe (1985) showed that the addition of a second, cold electron component to the dispersion equation could raise the frequency of wave growth to the vicinity of the electron plasma frequency. Ashour-Abdalla and Okuda (1986) further showed that, for sufficiently large cold electron density, the ion/ion relative drift drives an electron acoustic instability with characteristic frequency near the cold electron plasma frequency. The Ashour-Abdalla and Okuda electrostatic computer simulation of this instability exhibited stabilization by heating of the cold electrons, commensurate with the simulation results of Lin *et al.* (1985) for the electron/electron acoustic instability. More recent theory and simulations in this four-component model of the plasma sheet boundary layer have been carried out by Schriver and Ashour-Abdalla (1990), who also summarize the large amount of additional theoretical and computational studies concerning this region that have not been cited here.

Other theoretical studies of the ion/ion acoustic instability have been carried out in the contexts of the Earth's bow shock (Akimoto and Winske, 1985) and auroral field lines in the magnetosphere (Bergmann, 1984).

3.3 Electrostatic component/component instabilities in magnetized plasmas

The derivation of the linear dispersion equation for a plasma bearing a uniform zeroth-order magnetic field $\mathbf{B}_o = \hat{\mathbf{z}}B_o$ proceeds as in Section 2.3 through Equation (2.3.5). Thus, defining the jth component susceptibility as in Equation (2.2.5),

$$n_j^{(1)}(\mathbf{k},\omega) = -\frac{k^2\phi^{(1)}(\mathbf{k},\omega)}{4\pi e_j}K_j(\mathbf{k},\omega) \tag{3.3.1}$$

one obtains the linear electrostatic dispersion equation

$$1 + \sum_j K_j(\mathbf{k},\omega) = 0 \tag{3.3.2}$$

where, from Equation (2.3.5),

$$K_j(\mathbf{k},\omega) = -\frac{\omega_j^2}{n_j k^2}\int d^3v \int dt'\, i\mathbf{k}\cdot\frac{\partial f_j^{(0)}}{\partial \mathbf{v}'}\exp\{i[\mathbf{k}\cdot(\mathbf{x}'-\mathbf{x}) - \omega(t'-t)]\}. \tag{3.3.3}$$

As in Section 2.3 the unperturbed orbits are those given in Appendix B and the argument of the exponential is $ib_j(\tau,\omega)$ where b_j is given by Equation (2.3.8).

If the drifting Maxwellian distribution function of Equation (3.1.3) is used with $\mathbf{v}_{oj}$ parallel to the magnetic field $\mathbf{B}_o$, the derivation of the susceptibility proceeds much as in Section 2.3. Thus, following the first procedure of Appendix C,

$$K_j(\mathbf{k},\omega) = \frac{k_j^2}{k^2}\left[1 + \frac{(\omega - \mathbf{k}\cdot\mathbf{v}_{oj})\exp(-\lambda_j)}{\sqrt{2}|k_z|v_j}\sum_{m=-\infty}^{\infty} I_m(\lambda_j)Z(\zeta_j^m)\right] \tag{3.3.4}$$

where $\lambda_j \equiv (k_y a_j)^2$ and the argument of the plasma dispersion function is now

$$\zeta_j^m \equiv \frac{\omega - \mathbf{k}\cdot\mathbf{v}_{oj} + m\Omega_j}{\sqrt{2}|k_z|v_j}.$$

In the limit of parallel propagation $k_y = 0, k_z = k$ and Equation (3.3.4) reduces to Equation (3.2.3). Thus at $\mathbf{k}\times\mathbf{B}_o = 0$, linear dispersion reduces to that of an unmagnetized plasma and the results of Section 3.2 are recovered.

In the limit of perpendicular propagation, $k_z = 0$, so that $\mathbf{k}\cdot\mathbf{v}_{oj} = 0$ and the susceptibility reduces to that of Equation (2.3.11). Thus drift velocities parallel to the magnetic field introduce no new physics at strictly perpendicular propagation, and it is necessary only to consider oblique propagation here.

An alternate form of the jth component susceptibility in what is commonly called the Gordeyev integral form can be obtained by integrating Equation (3.3.3) only over the three velocity components. Then, again using the drifting Maxwellian distribution of Equation (3.1.3), one follows the second procedure outlined in Appendix C to obtain

$$K_j(\mathbf{k},\omega) = \frac{k_j^2}{k^2} + i\frac{k_j^2}{k^2}(\omega - \mathbf{k}\cdot\mathbf{v}_{oj})$$

$$\times \int_0^\infty d\tau\, \exp[i(\omega - \mathbf{k}\cdot\mathbf{v}_{oj})\tau]\exp(-k_z^2 v_j^2\tau^2/2)\exp[-2k_y^2 a_j^2 \sin^2(\Omega_j\tau/2)]. \tag{3.3.5}$$

The susceptibility reduces to its unmagnetized form under conditions that permit the small argument expansion of $\sin^2(\Omega_j\tau/2)$ in the third exponential factor of the integrand of the Gordeyev integral. There are two such distinct conditions on the wavevector:

$$k_y^2 a_j^2 \geq 9 \quad\text{and}\quad k_z^2 a_j^2 \geq 1/2 \tag{3.3.6}$$

(Gary, 1970) and

$$k_z^2 a_j^2 \geq 4 \tag{3.3.7}$$

(Gary, 1971). The oft-quoted condition $\omega_r \gg |\Omega_j|$ does not, by itself, reduce Equation (3.3.5) to the unmagnetized form, although this high frequency case often corresponds to the condition

$$\gamma \gg |\Omega_j|, \tag{3.3.8}$$

which does permit the unmagnetized expansion to be used. Because our procedure here is to assume a value for **k** and solve for ω, we do not know a priori whether Condition (3.3.8) is satisfied, and it is preferable to invoke Conditions (3.3.6) or (3.3.7) to validate use of the unmagnetized form of the susceptibility. Then, given (3.3.6) or (3.3.7), we obtain

$$K_j(\mathbf{k}, \omega) = -\frac{k_j^2}{2k^2} Z'\left(\frac{\omega - k_z v_{oj}}{\sqrt{2}\, k v_j}\right). \tag{3.3.9}$$

Problem 3.3.1. Show that both Equation (3.3.4) and, in the appropriate limit, Equation (3.3.9) follow from Equation (3.3.5).

3.3.1 The electron/ion cyclotron instability

In this subsection we consider the configuration discussed earlier in Subsection 3.2.1; that is, an electron-ion plasma in which both species consist of a single component. The presence of an ion-electron relative drift velocity at $\mathbf{v}_o \times \mathbf{B}_o = 0$ permits some of the ion cyclotron waves discussed in Section 2.3.1 to become unstable. As v_{oe} is increased from zero, the three more heavily damped oblique cyclotron modes illustrated in Figure 2.7 do not experience significant changes in either ω_r or γ. However, the mode that is least damped at $\theta \leq 87°$ undergoes reduced damping as the electron/ion relative drift velocity increases, and eventually attains $\gamma > 0$. Figure 3.18 traces the evolution of the dispersion of this mode as v_{oe} is increased from zero through and beyond the threshold of the electron/ion cyclotron instability (Drummond and Rosenbluth, 1962; Kindel and Kennel, 1971).

Because this instability propagates obliquely to $\mathbf{B}_o$, a concise and accurate expression for $\omega(\mathbf{k})$ is not available. The following properties are generally valid near threshold: $\gamma \ll \omega_r$, $\Omega_i < \omega_r < 2\Omega_i$, $k_z \ll k_y$ and $ka_i \sim 1$. The instability is driven by a strong Landau resonance on the electrons: $|\zeta_e^0| < 1$, but the ions are nonresonant with respect to this interaction: $|\zeta_i^0| \gg 1$. The electrons do not participate in the cyclotron resonances, because $|\zeta_e^{\pm 1}| \gg 1$; damping is due to a weak ion cyclotron resonance: $1 < |\zeta_i^{-1}| \lesssim 2$.

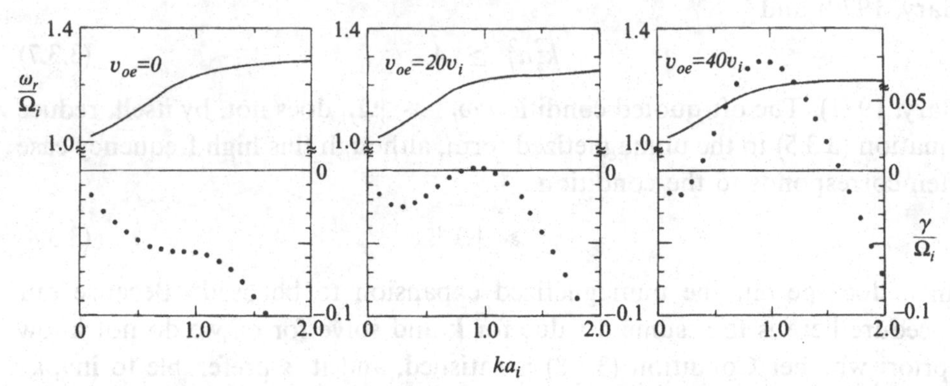

Fig. 3.18 The real frequencies (solid lines) and damping/growth rates (dotted lines) for the electron/ion cyclotron instability as functions of wavenumber for three different values of the electron/ion relative drift speed at $\theta = 85°$. Here, and in Figure 3.19, we consider an electron-proton plasma with $\omega_e = 10|\Omega_e|$, $v_{oi} = 0$, and $T_e = T_i$.

A comparison of thresholds for two electrostatic electron/ion instabilities is given in Figure 3.4. This figure demonstrates that the electron/ion cyclotron instability has a lower threshold than the electron/ion acoustic instability for $0.10 < T_e/T_i < 9.0$. In general, the θ value at threshold of this instability decreases with increasing T_e/T_i, and at $T_e/T_i \gtrsim 50$, the electron/ion cyclotron instability loses its identity by merging into the electron/ion acoustic instability.

As the drift speed is raised above threshold, successively higher cyclotron harmonic waves become unstable. Figure 3.19 illustrates the maximum growth rates of these harmonics as a function of v_{oe}; for the parameters considered here the higher harmonics have both successively higher thresholds and successively smaller growth rates.

Also plotted in Figure 3.19 is γ_m of the electron/ion acoustic instability for these parameters. By comparison with the ion cyclotron modes, the maximum growth of this mode increases very rapidly with electron/ion relative drift. Thus, once the electron/ion acoustic instability becomes unstable, it dominates the linear, and probably the nonlinear, physics and there is little reason to investigate the large v_{oe} limit of the ion cyclotron instability.

3.3.2 *The ion/ion cyclotron instability*

The presence of two ion components with a relative drift $\mathbf{v}_o \times \mathbf{B}_o = 0$ can also drive unstable the electrostatic ion cyclotron waves of Section 2.3.1. Given this configuration, Figure 3.20 shows the ion core/electron relative

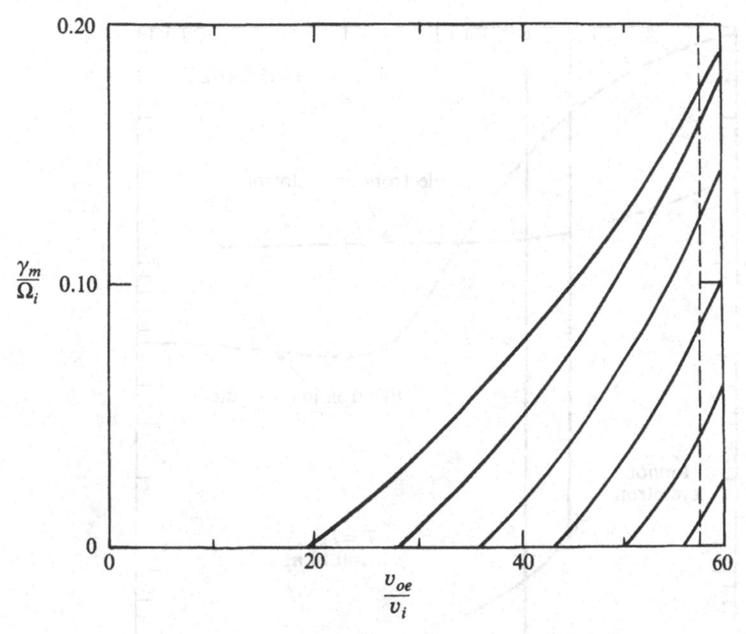

Fig. 3.19 The maximum growth rates of electrostatic electron/ion instabilities as functions of the electron/ion drift speed. The solid lines represent, from left to right, the $m = 1,...6$ harmonics of the electron/ion cyclotron instability; the dashed line represents the electron/ion acoustic instability.

drift speed at threshold of four growing modes: the ion/ion acoustic, the electron/ion acoustic, the electron/ion cyclotron, and the ion/ion cyclotron instabilities. Note that, like the ion/ion acoustic instability, the ion/ion cyclotron instability has a T_e/T_c cutoff below which the mode is stable for all drift speeds. However, because the ion/ion cyclotron mode can grow at somewhat smaller T_e/T_c ratios than the ion/ion acoustic instability, and because both ion/ion modes have much lower threshold drift speeds than the two electron/ion modes, there is a definite, albeit limited, range of parameter space in which the ion/ion cyclotron instability can be the only growing mode in this configuration.

The dispersion properties of the ion/ion cyclotron instability (see Figure 8.12) are similar to those of the electron/ion cyclotron instability; for the unstable mode of lowest threshold $1.0 < \omega_r/\Omega_i < 2.0$, $\gamma_m \ll \Omega_i$, $ka_i \simeq 1$, and $0° \ll \theta < 90°$. The parametric dependences of γ_m for this instability are often similar to those of the ion/ion acoustic instability: the maximum growth rate increases as the ion beam density increases and as the ion beam temperature decreases relative to the ion core temperature.

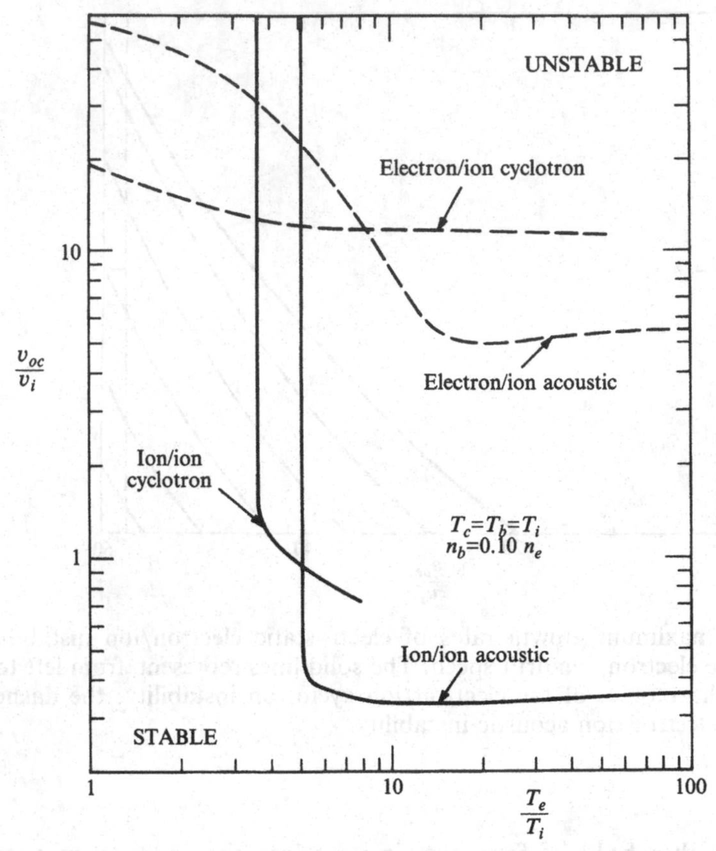

Fig. 3.20 The threshold ion core/electron drift speeds of four instabilities as functions of T_e/T_i. Here the plasma consists of three components: the electrons (e), a more dense ion core (c) and a less dense ion beam (b). Here $T_c = T_b = T_i$ and $n_b = 0.10 n_e$.

3.3.3 Applications of electrostatic ion cyclotron instabilities

Plasma modes that propagate obliquely to $\mathbf{B}_o$ generally involve fluctuating electric fields with components that are both longitudinal and transverse to $\mathbf{k}$. Thus the applicability of the electrostatic cyclotron instabilities discussed in the preceeding subsections is limited to plasmas that satisfy $\beta \ll 1$, such as the ionosphere and the low altitude magnetosphere. The primary applications of the electron/ion cyclotron instability have been in the auroral ionosphere (Kindel and Kennel, 1971), where this mode has been used to explain ion precipitation in the diffuse aurora (Ashour-Abdalla and Thorne, 1978).

Because both ion beams and electron-borne field-aligned currents have been observed on auroral field lines near one Earth radius, there has been a long-standing question as to which source of free energy is primarily responsible for the enhanced electric fluctuations observed between the harmonics of the proton cyclotron frequency (Cattell, 1981, and references therein). Kaufmann and Kintner (1982, 1984) showed that both the electron/ion cyclotron and the ion/ion cyclotron instabilities can arise along auroral field lines. The model of Kaufmann and Kintner has been refined by André (1985). Bergmann (1984) further showed that, although the ion/ion cyclotron instability grows only at $T_c < T_e$ and $T_b < T_e$, the electron/ion cyclotron instability may be unstable at the equal electron and ion component temperatures more characteristic of auroral zone plasmas. Thus Figure 3.20 confirms Bergmann's (1984) conclusion that the electron/ion cyclotron instability is the more likely source of enhanced proton cyclotron fluctuations observed near one Earth radius in the auroral zone.

Another issue of interest has been whether these or other fluctuations observed along auroral field lines can provide strong perpendicular heating of ions and thereby act as a source of the strongly anisotropic ion conic distributions also observed there. Okuda and Ashour-Abdalla (1983) and Ashour-Abdalla and Okuda (1984) used linear theory, quasilinear theory and electrostatic computer simulations to study the heating of hydrogen and heavy ions by the electron/ion cyclotron instability. In particular, Ashour-Abdalla and Okuda (1984) used a drift speed $v_{oe} = v_e$ well above the instability threshold and a model in which a constant flux of Maxwellian electrons enters at the ionospheric boundary of the simulation; their results show strong perpendicular heating of both protons and oxygen ions, and saturation of the proton cyclotron instability due to the quasilinear increase in the proton $T_{\perp}/T_{\parallel}$.

Kintner and Gorney (1984) searched the S3-3 data set for examples of perpendicular ion acceleration that were simultaneous with broadband electric field fluctuations. They found only one such event, during which there was no evidence of hydrogen cyclotron waves, though there were relatively weak amplitude fluctuations that may have corresponded to Doppler-broadened oxygen cyclotron waves. Furthermore, the electron drift velocity was too small to locally excite ion cyclotron instabilities. The absence of correlation between ion cyclotron fluctuations and strong ion heating, plus the requirement of a relatively large electron drift speed, have left the electron/ion cyclotron instability in some disfavor as a means for ion conic generation.

At higher altitudes in the auroral zone (above 5000 km), upwelling ion beams are observed that include streaming protons, helium ions and oxygen ions. These beams are believed to be accelerated by quasistatic electric fields associated with oblique double layers; because lighter ions gain the same energy but greater velocities in falling through the same potential drop as heavier ions, one may expect ion/ion instabilities to arise. Because $\omega_p \sim \Omega_p$ in the auroral zone, the distinction between ion acoustic and ion cyclotron instabilities is not as clear as when $\omega_p \gg \Omega_p$; however, because $\omega_r \sim \Omega_p$ and $k_y a_p \sim 1$, both instabilities are ion-cyclotron-like in character. Linear theory of oxygen-ion/proton electrostatic instabilities in the high altitude auroral region (Bergmann *et al.*, 1988; Dusenbery *et al.*, 1988) has demonstrated that the most unstable modes are usually obliquely propagating ion/ion instabilities. Winglee *et al.* (1989) assumed very hot electrons ($T_e = 100T_i$) as initial conditions in their electrostatic computer simulations; they obtained substantial heating of both the proton and heavy ion components through the growth of an electrostatic ion/ion instability. The computer simulations of Schriver *et al.* (1990) began with less hot electrons ($T_e \simeq 10T_i$) and showed that an electron/ion instability first heated the electrons, after which an ion/ion instability led to the heating of both proton and oxygen ion components.

3.4 Summary

The electrostatic instabilities of a collisionless plasma that have been discussed in this Chapter are summarized in Table 3.1. Virtually all space plasmas bear a magnetic field. In a plasma with $\beta \ll 1$, the background magnetic field energy is much larger than the plasma thermal energy, so that magnetic fluctuations do not easily couple to the thermal plasma fluctuations and the electrostatic approximation used here is valid. As β increases, the coupling between electric and magnetic fluctuations increases and the electrostatic approximation is no longer valid for the cyclotron instabilities of Table 3.1. However, the various "unmagnetized" instabilities such as the acoustic modes may propagate as electrostatic modes at or near $\mathbf{k} \times \mathbf{B}_o = 0$ in plasmas of arbitrary β.

At relative drift speeds at or somewhat above threshold, the component/component instabilities described in this chapter are resonant. That is, they correspond to $|\zeta_j| \lesssim 1$ (or, in the case of the cyclotron instabilities, to $|\zeta_j^{\pm 1}| \lesssim 1$) for at least one component j and require a full kinetic description. As the relative drift speed increases to well above the thermal speeds of the components, the electron/electron and electron/ion instabilities can become nonresonant modes for which $|\zeta_j| \gg 1$ for all components j; such modes are usually well described by the cold plasma approximation.

Table 3.1. *Some electrostatic component/component plasma instabilities*

Name	Source	Frequency	Wavevector	Conditions
		Resonant instabilities		
Langmuir beam	Electron beam-core drift	$\omega_r^2 \simeq \omega_e^2 + 3k^2 v_e^2$	$k \ll k_e$	$n_b/n_e < (v_b/v_{ob})^3$
Electron/electron beam	Electron beam-core drift	$\omega_r \simeq k_z v_{ob}$	$k \ll k_e$	$\frac{n_b}{n_e} \gtrsim \left(\frac{v_b}{v_{ob}}\right)^3$
Electron/electron acoustic	Electron hot-cool drift	$\omega_r \simeq \omega_c$	$k_h < k < k_c$	$T_c \ll T_h$
Electron/ion acoustic	Electron-ion drift	$\omega_r \simeq k \left(\frac{T_e+3T_i}{m_i}\right)^{1/2}$	$k < k_e$	$T_i \ll T_e$, $c_s < v_o$
Ion/ion acoustic	Ion beam-core drift	$\omega_r \simeq k \left(\frac{T_e+3T_i}{m_i}\right)^{1/2}$	$k < k_e$	$T_i \ll T_e$, $v_i \lesssim v_o$
Electron/ion cyclotron	Electron-ion drift $\mathbf{v}_o \times \mathbf{B}_o = 0$	$\Omega_i < \omega_r < 2\Omega_i$, $2\Omega_i < \omega_r < 3\Omega_i$, etc.	$ka_i \sim 1$, $0 < k_z \ll k_y$	$0.1 \lesssim T_e/T_i \lesssim 9$ $c_s < v_o$
Ion/ion cyclotron	Ion beam-core drift $\mathbf{v}_o \times \mathbf{B}_o = 0$	$\Omega_i < \omega_r < 2\Omega_i$, $2\Omega_i < \omega_r < 3\Omega_i$, etc.	$ka_i \sim 1$, $0 < k_z \ll k_y$	$T_i < T_e$ $v_i < v_o$
		B. Nonresonant instabilities		
Electron/electron two-stream	Electron beam-core drift	$\omega_r < \omega_e$	$k \simeq \omega_e/v_{ob}$	$v_b, v_c \ll v_o$
Electron/ion two-stream (Buneman)	Electron-ion drift	$\omega_r \sim \omega_i$	$k \simeq \omega_e/v_o$	$v_i, v_e \ll v_o$

References

Abas, S., and S. P. Gary, Ion wave instability in an electrostatic shock, *Plasma Phys.*, **13**, 262, 1971.

Akimoto, K., and N. Omidi, The generation of broadband electrostatic noise by an ion beam in the magnetotail, *Geophys. Res. Lett.*, **13**, 97, 1986.

Akimoto, K., and D. Winske, Ion-acoustic-like waves excited by the reflected ions at the Earth's bow shock, *J. Geophys. Res.*, **90**, 12095, 1985.

Anderson, R. R., G. K. Parks, T. E. Eastman, D. A. Gurnett and L. A. Frank, Plasma waves associated with energetic particles streaming into the solar wind from the Earth's bow shock, *J. Geophys. Res.*, **86**, 4493, 1981.

André, M., Ion waves generated by streaming particles, *Annales Geophys.*, **3**, 1, 1985.

Ashour-Abdalla, M., and H. Okuda, Turbulent heating of heavy ions on auroral field lines, *J. Geophys. Res.*, **89**, 2235, 1984.

Ashour-Abdalla, M., and H. Okuda, Electron acoustic instabilities in the geomagnetic tail, *Geophys. Res. Lett.*, **13**, 366, 1986.

Ashour-Abdalla, M., and R. M. Thorne, Toward a unified view of diffuse auroral precipitation, *J. Geophys. Res.*, **83**, 4755, 1978.

Bergmann, R., Electrostatic ion (hydrogen) cyclotron and ion acoustic wave instabilities in regions of upward field-aligned current and upward ion beams, *J. Geophys. Res.*, **89**, 953, 1984.

Bergmann, R., I. Roth and M. K. Hudson, Linear stability of the H^+-O^+ two-stream interaction in a magnetized plasma, *J. Geophys. Res.*, **93**, 4005, 1988.

Buneman, O., Instability, turbulence and conductivity in current-carrying plasma, *Phys. Rev.*, **115**, 503, 1959.

Burch, J. L., P. H. Reiff and M. Sugiura, Upward electron beams measured by DE-1: A primary source of dayside region 1 Birkeland currents, *Geophys. Res. Lett.*, **10**, 753, 1983.

Cattell, C., The relationship of field-aligned currents to electrostatic ion cyclotron waves, *J. Geophys. Res.*, **86**, 3641, 1981.

Drummond, W. E., and M. N. Rosenbluth, Anomalous diffusion arising from microinstabilities in a plasma, *Phys. Fluids*, **5**, 1507, 1962.

Dum, C. T., E. Marsch and W. Pilipp, Determination of wave growth from measured distribution functions and transport theory, *J. Plasma Phys.*, **23**, 91, 1980.

Dusenbery, P. B., R. F. Martin, Jr. and R. M. Winglee, Ion-ion waves in the auroral region: wave excitation and ion heating, *J. Geophys. Res.*, **93**, 5655, 1988.

Etcheto, J., and M. Faucheux, Detailed study of electron plasma waves upstream of the Earth's bow shock, *J. Geophys. Res.*, **89**, 6631, 1984.

Feldman, W. C., R. C. Anderson, S. J. Bame, S. P. Gary, J. T. Gosling, D. J. McComas, M. F. Thomsen, G. Paschmann and M. M. Hoppe, Electron velocity distributions near the Earth's bow shock, *J. Geophys. Res.*, **88**, 96, 1983.

Filbert, P. C., and P. J. Kellogg, Electrostatic noise at the plasma frequency beyond the Earth's bow shock, *J. Geophys. Res.*, **84**, 1369, 1979.

Fitzenreiter, R. J., A. J. Klimas and J. D. Scudder, Detection of bump-on-tail reduced electron velocity distributions at the electron foreshock boundary, *Geophys. Res. Lett.*, **11**, 496, 1984.

Fredricks, R. W., C. F. Kennel, F. L. Scarf, C. R. Crook and I. M. Green, Detection of electric field disturbance in the Earth's bow shock, *Phys. Rev. Lett.*, **21**, 1761, 1968.

Fried, B. D., and R. W. Gould, Longitudinal ion oscillations in a hot plasma, *Phys. Fluids*, **4**, 139, 1961.

Fried, B. D., and A. Y. Wong, Stability limits for longitudinal waves in ion beam-plasma interaction, *Phys. Fluids*, **9**, 1084, 1966.

Fried, B. D., and A. Y. Wong, Reply to comments by Kaya Imre and Ercument Ozizmir, *Phys. Fluids*, **17**, 1048, 1974.

Fuselier, S. A., and D. A. Gurnett, Short wavelength ion waves upstream of the Earth's bow shock, *J. Geophys. Res.*, **89**, 91, 1984.

Fuselier, S. A., D. A. Gurnett and R. J. Fitzenreiter, The downshift of electron plasma oscillations in the electron foreshock region, *J. Geophys. Res.*, **90**, 3935, 1985.

Gary, S. P., Longitudinal waves in a perpendicular collisionless plasma shock, II. Vlasov ions, *J. Plasma Phys.*, **4**, 753, 1970.

Gary, S. P., Longitudinal waves in a perpendicular collisionless plasma shock, III. $T_e \sim T_i$, *J. Plasma Phys.*, **6**, 561, 1971.

Gary, S. P., Ion-acoustic-like instabilities in the solar wind, *J. Geophys. Res.*, **83**, 2504, 1978.

Gary, S. P., Electrostatic instabilities in plasmas with two electron components, *J. Geophys. Res.*, **90**, 8213, 1985.

Gary, S. P., The electron/electron acoustic instability, *Phys. Fluids*, **30**, 2745, 1987.

Gary, S. P. and N. Omidi, The ion/ion acoustic instability, *J. Plasma Phys.*, **37**,45, 1987.

Gosling, J. T., M. F. Thomsen, S. J. Bame, T. G. Onsager and C. T. Russell, The electron edge of the low latitude boundary layer during accelerated flow events, *Geophys. Res. Lett.*, **17**, 1833, 1990.

Grabbe, C. L., New results on the generation of broadband electrostatic waves in the magnetotail, *Geophys. Res. Lett.*, **12**, 483, 1985.

Grabbe, C. L., and T. E. Eastman, Generation of broadband electrostatic noise by ion beam instabilities in the magnetotail, *J. Geophys. Res.*, **89**, 3865, 1984.

Greenstadt, E. W., and M. M. Mellott, Plasma wave evidence for reflected ions in front of subcritical shocks: ISEE 1 and 2 observations, *J. Geophys. Res.*, **92**, 4730, 1987.

Gurnett, D. A., Plasma waves and instabilities, in *Collisionless Shocks in the Heliosphere: Reviews of Current Research*, Geophysical Monograph 35, American Geophysical Union, Washington, DC, 1985.

Gurnett, D. A., Waves and instabilities, in *Physics of the Inner Heliosphere II*, R. Schwenn and E. Marsch, editors, Springer-Verlag, Berlin, 1991.

Gurnett, D. A., L. A. Frank and R. Lepping, Plasma waves in the distant magnetotail, *J. Geophys. Res.*, **81**, 6059, 1976.

Imre, K., and E. Ozizmir, Comments on "Stability limits for longitudinal waves in ion beam-plasma interaction," *Phys. Fluids*, **17**, 1046, 1974.

Kaufmann, R. L., and P. M. Kintner, Upgoing ion beams: 1. Microscopic analysis, *J. Geophys. Res.*, **87**, 10487, 1982.

Kaufmann, R. L., and P. M. Kintner, Upgoing ion beams: 2. Fluid analysis and magnetosphere-ionosphere coupling, *J. Geophys. Res.*, **89**, 2195, 1984.

Kindel, J. M., and C. F. Kennel, Topside current instabilities, *J. Geophys. Res.*, **76**, 3055, 1971.

Kintner, P. M., and D. J. Gorney, A search for the plasma processes associated with perpendicular heating, *J. Geophys. Res.*, **89**, 937, 1984.

Klimas, A. J., A mechanism for plasma waves at the harmonics of the plasma frequency in the electron foreshock boundary, *J. Geophys. Res.*, **88**, 9081, 1983.

Kurth, W. S., D. A. Gurnett and F. L. Scarf, High-resolution spectrograms of ion acoustic waves in the solar wind, *J. Geophys. Res.*, **84**, 3413, 1979.

Lacombe, C., A. Mangeney, C. C. Harvey and J. D. Scudder, Electron plasma waves upstream of the Earth's bow shock, *J. Geophys. Res.*, **90**, 73, 1985.

Lemons, D. S., J. R. Asbridge, S. J. Bame, W. C. Feldman, S. P. Gary and J. T. Gosling, The source of electrostatic fluctuations in the solar wind, *J. Geophys. Res.*, **84**, 2135, 1979.

Lin, C. S., J. L. Burch, S. D. Shawhan and D. A. Gurnett, Correlation of auroral hiss and upward electron beams near the polar cusp, *J. Geophys. Res.*, **89**, 925, 1984.

Lin, C. S., D. Winske and R. L. Tokar, Simulation of the electron acoustic instability in the polar cusp, *J. Geophys. Res.*, **90**, 8269, 1985.

Marsch, E., Beam-driven electron acoustic waves upstream of the Earth's bow shock, *J. Geophys. Res.*, **90**, 6327, 1985.
Marsch, E., Kinetic physics of the solar wind plasma, in *Physics of the Inner Heliosphere II*, R. Schwenn and E. Marsch, editors, Springer-Verlag, Berlin, 1991.
Okuda, H., and M. Ashour-Abdalla, Acceleration of hydrogen ions and conic formation along auroral field lines, *J. Geophys. Res.*, **88**, 899, 1983.
Omidi, N., Broadband electrostatic noise produced by ion beams in the Earth's magnetotail, *J. Geophys. Res.*, **90**, 12330, 1985.
O'Neil, T. M., and J. H. Malmberg, Transition of the dispersion roots from beam-type to Landau-type solutions, *Phys. Fluids*, **11**, 1754, 1968.
Onsager, T. G., and R. H. Holzworth, Measurement of the electron beam mode in Earth's foreshock, *J. Geophys. Res.*, **95**, 4175, 1990.
Onsager, T. G., M. F. Thomsen, J. T. Gosling and S. J. Bame, Electron distributions in the plasma sheet boundary layer: time-of-flight effects, *Geophys. Res. Lett.*, **17**, 1837, 1990.
Scarf, F. L., R. W. Fredricks, L. A. Frank and M. Neugebauer, Nonthermal electrons and high-frequency waves in the upstream solar wind, 1, Observations, *J. Geophys. Res.*, **76**, 5162, 1971.
Scarf, F. L., L. A. Frank, K. L. Ackerson and R. P. Lepping, Plasma wave turbulence at distant crossings of the plasma sheet boundaries and the neutral sheet, *Geophys. Res. Lett.*, **1**, 189, 1974.
Schriver, D., and M. Ashour-Abdalla, Cold plasma heating in the plasma sheet boundary layer: theory and simulations, *J. Geophys. Res.*, **95**, 3987, 1990.
Schriver, D., M. Ashour-Abdalla, H. Collin and N. Lallande, Ion beam heating in the auroral zone, *J. Geophys. Res.*, **95**, 1015, 1990.
Thomsen, M. F., H. C. Barr, S. P. Gary, W. C. Feldman and T. E. Cole, Stability of electron distributions within the Earth's bow shock, *J. Geophys. Res.*, **88**, 3035, 1983.
Tokar, R. L., and S. P. Gary, Electrostatic hiss and the beam driven electron acoustic instability in the dayside polar cusp, *Geophys. Res. Lett.*, **11**, 1180, 1984.
Winglee, R. M., P. B. Dusenbery, H. L. Collin, C. S. Lin and A. M. Persoon, Simulations and observations of heating of auroral ion beams, *J. Geophys. Res.*, **94**, 8943, 1989.

4

Electrostatic drift instabilities in inhomogeneous plasmas

Every plasma is inhomogeneous to some extent, and the associated plasma gradients are sources of free energy that can drive plasma instabilities. In this chapter we consider the linear theory of drift instabilities, modes driven unstable by a plasma gradient perpendicular to $\mathbf{B}_o$.

In the direction parallel to a magnetic field, pressure gradients give rise to electric fields that lead to currents and bulk plasma motion; that is, such gradients do not correspond to a steady-state description under a macroscopic description of the plasma. However, pressure gradients perpendicular to a magnetic field can correspond to a steady-state situation; that is, ∇P in the momentum equation of a one-fluid description of the plasma can be balanced by the $\mathbf{J} \times \mathbf{B}_o/c$ term. Nevertheless, such gradients do not correspond to an equilibrium plasma configuration; the zeroth-order distribution functions are non-Maxwellian and lead to the growth of plasma instabilities which act to dissipate the gradients.

In this chapter we consider, as before, collisionless plasmas with a uniform zeroth-order magnetic field $\mathbf{B}_o = \hat{\mathbf{z}}B_o$. In Section 4.1 we discuss a model distribution function for density gradients perpendicular to a uniform magnetic field, examine the associated linear dispersion equation and discuss the two most popular density drift instabilities. Section 4.2 describes the instability properties that result when a plasma with a density gradient is subject to a uniform acceleration, and Section 4.3 briefly summarizes some properties of temperature drift instabilities.

4.1 Density drift instabilities

We assume a weak density gradient perpendicular to $\mathbf{B}_o$ of the form

$$n_j^{(0)}(x) = n_j(1 + \epsilon_n x) \tag{4.1.1}$$

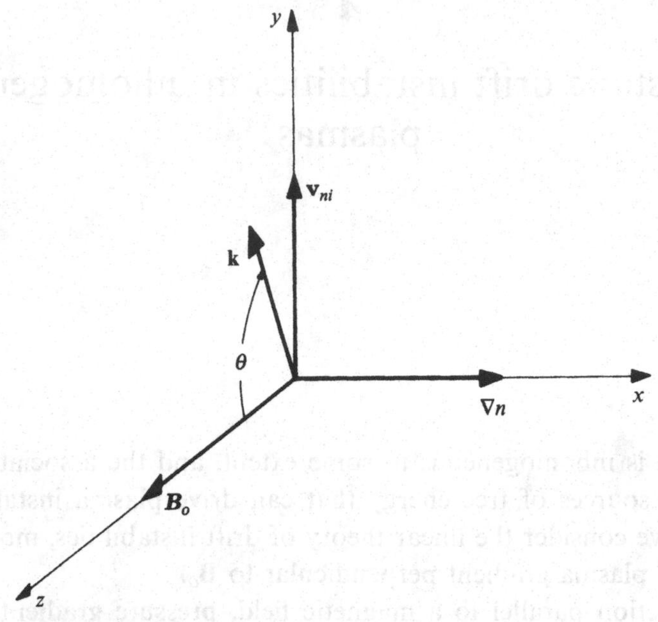

Fig. 4.1 The relationship among the magnetic field, the wavevector, the density gradient and the ion density gradient drift velocity.

where $\epsilon_n a_j \ll 1, |\epsilon_n x| \ll 1$, and $n_i = n_e$. Using constants of the motion (B.7) through (B.9), we construct a zeroth-order distribution function that satisfies the time-independent Vlasov equation and that yields (4.1.1):

$$f_j^{(0)}(x, \mathbf{v}) = \frac{n_j}{(2\pi v_j^2)^{3/2}} \left[1 + \epsilon_n (x + \frac{v_y}{\Omega_j}) \right] \exp(-v^2/2v_j^2). \qquad (4.1.2)$$

The first and second velocity moments of this distribution yield

$$\mathbf{\Gamma}_j^{(0)} = n_j v_{nj} \hat{\mathbf{y}}$$

$$n_j T_j = n_j T_j (1 + \epsilon_n x) + O(\epsilon_n^2)$$

where the density gradient drift speed of the jth species is

$$v_{nj} \equiv \frac{\epsilon_n v_j^2}{\Omega_j}.$$

Figure 4.1 shows the relationship among the various vector quantities in this configuration.

Because v_{nj} bears the sign of e_j, there is a nonzero current $\mathbf{J}_o$ associated with these drifts; this current represents a free energy that can drive instabilities. Ampére's equation (1.2.6) gives the gradient in $\mathbf{B}_o$ associated with this current. If $\beta = 8\pi n(T_e + T_i)/B_o^2$ is small, $|\nabla B_o(x)|/B_o = \epsilon_B \ll \epsilon_n$ and we may take the magnetic field to be uniform. Note that v_{nj} is independent of m_j; this implies that at $T_e \sim T_i$, $|v_{ne}| \sim v_{ni}$ and both electron and ion density drift velocities can contribute to instability growth. In fact, as we shall show, there are two classes of density drift instabilities, one driven by v_{ne} and one by v_{ni}.

To derive a linear dispersion equation appropriate for instabilities driven by density gradient drift velocities, we solve the linear Vlasov equation in the electrostatic approximation by the method of unperturbed orbits to obtain, as in Equation (2.3.2),

$$f_j^{(1)}(\mathbf{x}, \mathbf{v}, t) = \frac{e_j}{m_j} \int_{-\infty}^{t} dt' \, \nabla' \phi^{(1)}(\mathbf{x}', t') \cdot \frac{\partial f_j^{(0)}(x', \mathbf{v}')}{\partial \mathbf{v}'} \tag{4.1.3}$$

where a prime denotes a variable along the unperturbed orbit of a jth species particle and the fluctuations are assumed to be growing in time so that the contribution to $f_j^{(1)}$ from $t = -\infty$ may be ignored.

Because the model assumes a spatially homogeneous plasma in the y and z directions, one may Fourier transform all fluctuating quantities in those two directions and write

$$h^{(1)}(\mathbf{x}, t) = h^{(1)}(x, \mathbf{k}, \omega) \exp[i(\mathbf{k} \cdot \mathbf{x} - \omega t)] \tag{4.1.4}$$

where $\mathbf{k} = \hat{\mathbf{y}} k_y + \hat{\mathbf{z}} k_z$ and $h^{(1)}$ represents any first-order quantity. However, one may not Fourier transform in the direction of the inhomogeneity, which, in the slab model used here, is taken to be the x-direction. The traditional procedure for deriving a linear dispersion equation in the case of weak inhomogeneities has been to use the "local approximation" (Krall and Trivelpiece, 1973), in which the x-dependence of the fluctuating potential is ignored. If (4.1.3) is used in Poisson's equation with this approximation, a linear dispersion equation may be derived that, by virtue of the spatial dependence of $f_j^{(0)}$, is still a function of x. If $\epsilon_n(x) \equiv \frac{1}{n(x)} \frac{dn(x)}{dx}$ has a maximum at x_m, one may estimate the maximum growth rate by evaluating the dispersion equation at $x = x_m$. If there is no maximum in $\epsilon_n(x)$, as in the case of Equation (4.1.1), the procedure most often used is to evaluate the dispersion equation at $x = 0$.

Thomsen and Gary (1982) have carried out a detailed numerical evaluation of a nonlocal theory and have confirmed the general conclusion that the local theory yields an accurate estimate of the maximum growth rate as long as the condition

$$(\epsilon_n)_{max} \ll k$$

is satisfied. Here $(\epsilon_n)_{max}$ is the maximum inverse density gradient scale length on $n(x)$.

Here we use what we call the "modified local approximation." We assume

$$\phi^{(1)}(x, \mathbf{k}, \omega) = (1 - \epsilon_n x)\phi^{(1)}(\mathbf{k}, \omega). \tag{4.1.5}$$

We then derive the linear susceptibility of the jth species which, under these assumptions, turns out to be independent of x. Thus we no longer need to make the arbitrary assumption that the dispersion equation is to be evaluated at $x = 0$. In any event, the result is the same for the two forms of the approximation, and yields a valid approximation to the maximum growth rate as long as $\epsilon_n \ll k$.

Substituting (4.1.4) and (4.1.5) into (4.1.3), using Poisson's equation and defining the jth species susceptibility K_j by

$$n_j^{(1)}(x, \mathbf{k}, \omega) = -\frac{k^2\phi^{(1)}(\mathbf{k}, \omega)}{4\pi e_j} K_j(x, \mathbf{k}, \omega), \tag{4.1.6}$$

the linear dispersion equation is

$$1 - \epsilon_n x + \sum_j K_j(x, \mathbf{k}, \omega) = 0 \tag{4.1.7}$$

where

$$K_j(x, \mathbf{k}, \omega) = -\frac{\omega_j^2}{n_j k^2} \int d^3v \int dt' \; [-\hat{\mathbf{x}}\epsilon_n + i\mathbf{k}(1 - \epsilon_n x')] \cdot \frac{\partial f_j^{(0)}}{\partial \mathbf{v}'}$$

$$\exp\{i[\mathbf{k} \cdot (\mathbf{x}' - \mathbf{x}) - \omega(t' - t)]\}. \tag{4.1.8}$$

The unperturbed orbits are once again the same as those used in Section 2.3, so that the argument of the exponential is $ib_j(\tau, \omega)$ where b_j is given by Equation (2.3.8). Because the zeroth-order distribution function, Equation (4.1.2), contains the Maxwellian distribution as a factor, we may once again use the technique discussed in Appendix C to evaluate the susceptibility integrals; ignoring all terms of order ϵ_n^2, we obtain

$$K_j(x, \mathbf{k}, \omega) = \frac{k_j^2}{k^2} \left[1 + \frac{\omega - k_y v_{nj}}{\sqrt{2}|k_z| v_j} \exp(-\lambda_j) \sum_{m=-\infty}^{\infty} I_m(\lambda_j) Z(\zeta_j^m) \right.$$

$$\left. - \frac{\omega}{\sqrt{2}|k_z| v_j} \frac{\epsilon_n}{k_y} \exp(-\lambda_j) \sum_{m=-\infty}^{\infty} m I_m(\lambda_j) Z(\zeta_j^m) \right] \tag{4.1.9}$$

which, unlike the susceptibility derived from the conventional local approximation, is independent of x. Here $\lambda_j \equiv k_y^2 a_j^2$ and

$$\zeta_j^m \equiv \frac{\omega + m\Omega_j}{\sqrt{2}|k_z|v_j}.$$

For those density drift instabilities that satisfy $k^2 \ll k_j^2$, it follows from Equation (4.1.7) that the linear electrostatic dispersion equation is independent of x.

Problem 4.1.1. Derive Equation (4.1.9).

Therefore, under the assumption of a uniform density gradient, the modified local approximation yields a result that is valid for all x such that $|\epsilon_n x| \ll 1$ and also yields the same result as the conventional local approximation evaluated at $x = 0$.

In the limit of perpendicular propagation, $k_z = 0$, and by the asymptotic expansion of the plasma dispersion function,

$$K_j(\mathbf{k}, \omega) = \frac{k_j^2}{k^2}\left[1 - (\omega - k_y v_{nj})\exp(-\lambda_j)\sum_{m=-\infty}^{\infty}\frac{I_m(\lambda_j)}{\omega + m\Omega_j} + \frac{\omega v_{nj}}{k_y v_j^2}\exp(-\lambda_j)\sum_{m=-\infty}^{\infty}\frac{m\Omega_j I_m(\lambda_j)}{\omega + m\Omega_j}\right]. \quad (4.1.10)$$

4.1.1 The universal drift instability

The universal drift instability is driven by the free energy associated with the electron density gradient drift speed, v_{ne}. This permits a drift wave, $\omega \simeq k_y v_{ne}$, $0 < k_z \ll k_y$, to exist in the plasma. This mode can arise in density gradients that are very broad compared to an ion gyroradius, so $v_{ne} \ll v_i$; because for this instability $k_y a_i \lesssim 1$, it follows that $\omega_r \ll \Omega_i$.

The evolution of this instability as the density gradient drift speed increases is shown in Figure 4.2, where θ corresponds to γ_m in each case. At $ka_i \ll 1$, the mode well satisfies the drift wave dispersion relation $\omega = k_y v_{ne}$, but the growth rate is relatively small at these wavenumbers. At wavenumbers near maximum growth ($k_m a_i \simeq \pi^{-1/2}(T_i/T_e)^{1/3}$ for $1 \lesssim T_e/T_i \lesssim 10$), the mode is strongly dispersive. Beyond maximum growth, γ decreases slowly with wavenumber and growth may persist to $k_y a_i \gg 1$. However, relatively weak collisions can damp the universal at short wavelengths; see Gary *et al.* (1983).

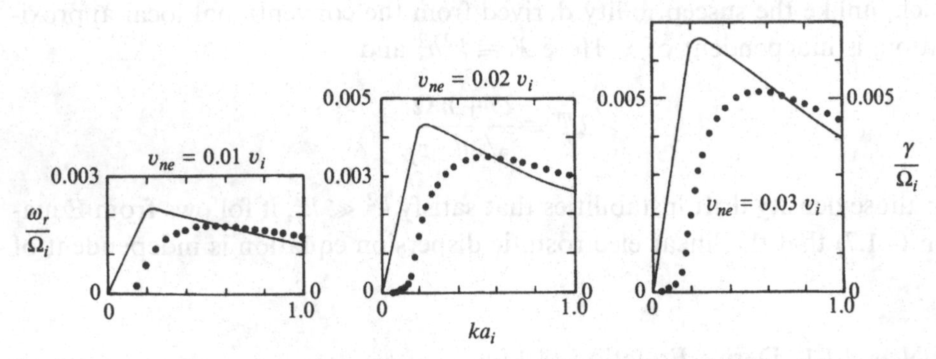

Fig. 4.2 The real frequency (solid line) and growth rate (dotted line) of the universal drift instability as functions of wavenumber for three different density drift speeds. In each case, θ is that of maximum growth rate; at $v_{ne} = 0.01v_i, \theta = 89.9935°$; at $v_{ne} = 0.02v_i, \theta = 89.987°$ and at $v_{ne} = 0.03v_i, \theta = 89.98°$. Here, and for all the figures in this chapter except as noted, $m_i = 1836m_e$, $T_e = T_i$, $\omega_e = 10|\Omega_e|$ and $\beta = 0$.

Because $\omega_r \ll \Omega_i$ for the universal, the $m = 0$ terms in Equation (4.1.9) make the most important contributions to this instability. Numerical evaluation of the linear dispersion equation demonstrates that, at $\epsilon_n a_i \ll 1, |m| \geq 1$ terms may be neglected for this mode. For this instability, the electrons are Landau resonant ($|\zeta_e^0| \lesssim 1$), but the ions are nonresonant: $\zeta_i^0 \gg 1$. Because $\omega_r \simeq \gamma$ near maximum growth, we have found no accurate analytic expression for the real frequency or growth rate in this case. Near maximum growth $\omega_r/k_y \simeq v_{ne}/3$ is approximately valid for $1 \lesssim T_e/T_i \lesssim 10$. And at long wavelengths, where $\gamma \ll \omega_r$, there is (Gary and Sanderson, 1978)

$$\omega_r \simeq \frac{k_y v_{ne} \exp(-\lambda_i) I_0(\lambda_i)}{1 + (T_e/T_i)[1 - \exp(-\lambda_i) I_0(\lambda_i)]}. \tag{4.1.11}$$

Problem 4.1.2. Derive Equation (4.1.11).

In the model used here, the universal has no threshold; that is, γ_m is proportional to v_{ne}. And, as $\epsilon_n a_i$ increases, higher frequency drift instabilities eventually emerge with much larger growth rates. So we do not present either threshold or asymptotic analyses of the universal here (however, see Figure 4.5).

As β increases from zero, the maximum growth of the universal drift instability is decreased, and at relatively modest values of this parameter the mode is stabilized ($\beta \simeq 0.15$ at $T_e = T_i$; see Berk and Dominguez, 1977; Huba and Gary, 1982).

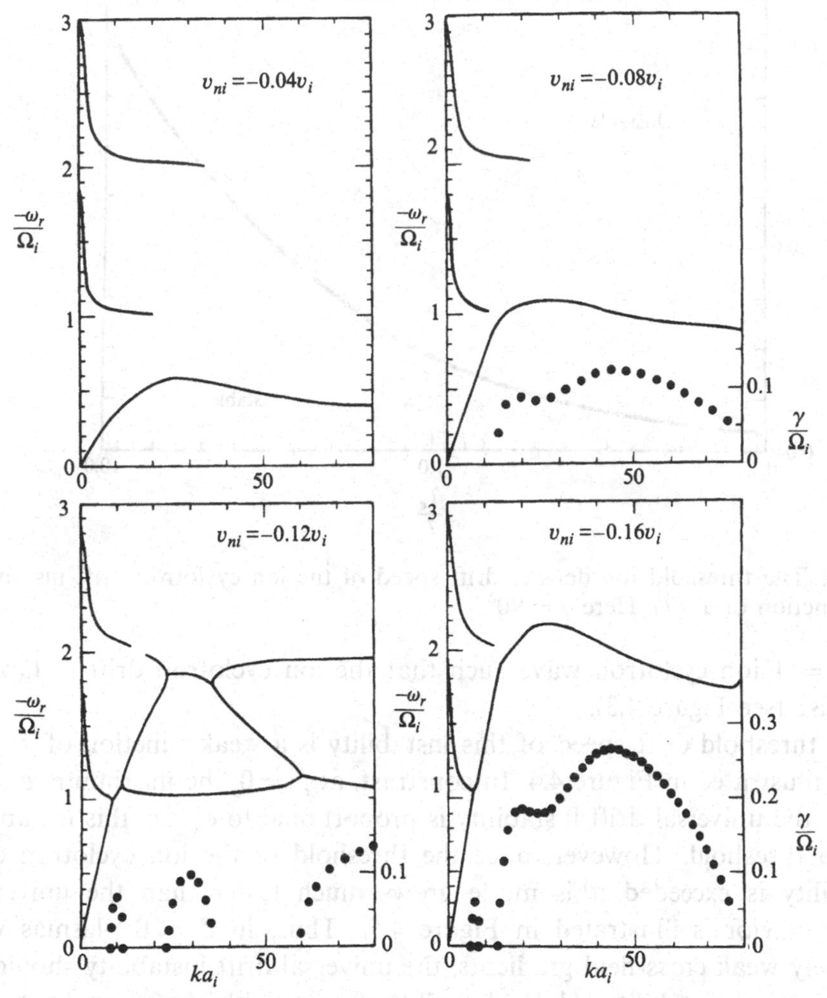

Fig. 4.3 The real frequency (solid lines) and growth rate (dotted lines) of the ion drift wave, the $m = 1$ and $m = 2$ ion cyclotron waves, and the ion cyclotron drift instability as functions of wavenumber for four different density drift speeds. Here $\theta = 90°$.

4.1.2 Ion cyclotron and lower hybrid drift instabilities

We now consider instabilities driven by the ion density drift. The ion density drift wave satisfies $\omega_r \simeq k_y v_{ni}$ at $k_y a_i \lesssim 1$, but has substantial dispersion at $k_y a_i \gg 1$, so that at sufficiently small v_{ni}/v_i, ω_r remains less than Ω_i for all ka_i. At perpendicular propagation there are also Bernstein or ion cyclotron waves at $\omega_r > \Omega_i$ (see Subsection 2.3.2). As the drift speed is increased, the frequency of the drift wave increases until it reaches the ion cyclotron frequency; at this point there is an interaction between the drift wave and

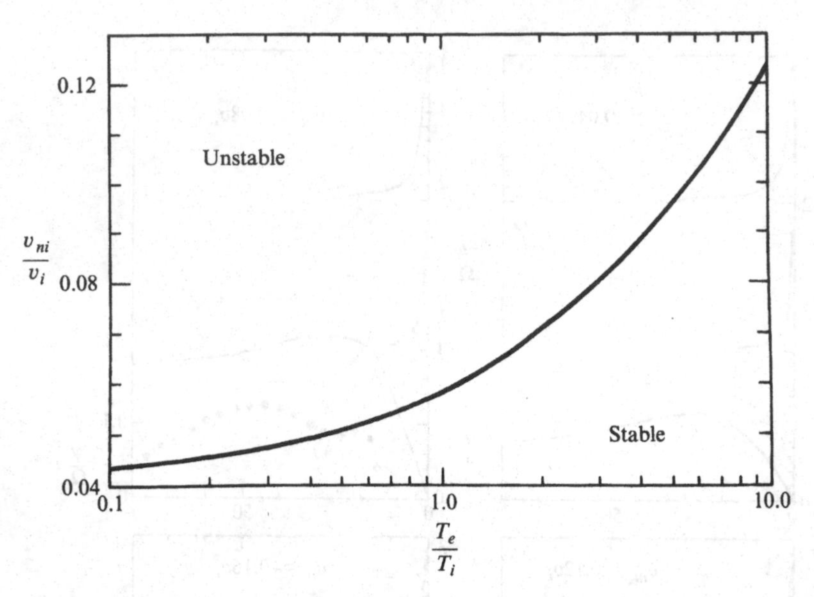

Fig. 4.4 The threshold ion density drift speed of the ion cyclotron drift instability as a function of T_e/T_i. Here $\theta = 90°$.

the $m = 1$ ion cyclotron wave such that the ion cyclotron drift instability can arise (see Figure 4.3).

The threshold drift speed of this instability is a weak function of T_e/T_i; this is illustrated in Figure 4.4. In constrast, at $\beta = 0$, the maximum growth rate of the universal drift instability is proportional to ϵ_n; i.e. this instability has no threshold. However, once the threshold of the ion cyclotron drift instability is exceeded, this mode grows much faster than the universal. This behavior is illustrated in Figure 4.5. Thus, in $\beta = 0$ plasmas with relatively weak cross-field gradients, the universal drift instability should be the dominant instability, while in similar plasmas with strong gradients, the ion cyclotron drift instability will grow much faster and probably dominate the microphysics.

As the drift speed is further increased, Figure 4.3 shows that the ion drift mode interacts with successively higher harmonic cyclotron waves and yields successively larger growth rates. When $|v_{ni}|/v_i$ becomes so large that $\gamma > \Omega_i$, the ion response to the fluctuating fields becomes unmagnetized. Then the ion susceptibility reduces to (Gary, 1980)

$$K_i(\mathbf{k}, \omega) = \frac{k_i^2}{k^2}\left[1 + \frac{(\omega - \mathbf{k} \cdot \mathbf{v}_{nj})}{\sqrt{2}kv_i} Z\left(\frac{\omega}{\sqrt{2}kv_i}\right)\right].$$

In this regime the mode is called the lower hybrid drift instability. In effect it is driven unstable by inverse Landau damping on the positive slope of

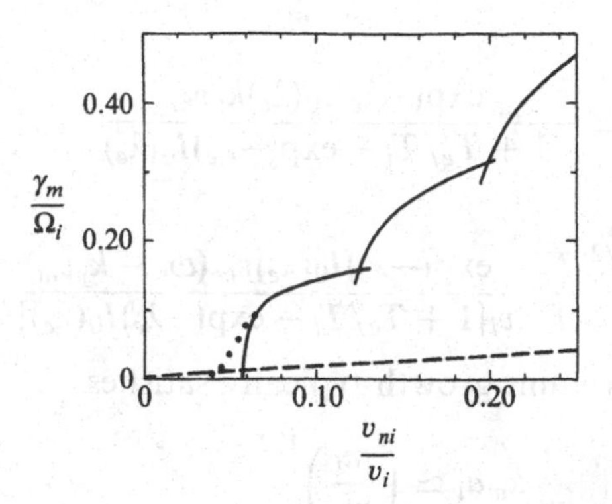

Fig. 4.5 The maximum growth rate of the universal and ion cyclotron density drift instabilities as functions of v_{ni} for $T_e = T_i$. The solid line represents the ion cyclotron instability at $\theta = 90°$; the dotted line is the same mode at $89.98° < \theta < 90°$, and the dashed line is the universal instability at $89.8° \leq \theta \leq 90°$.

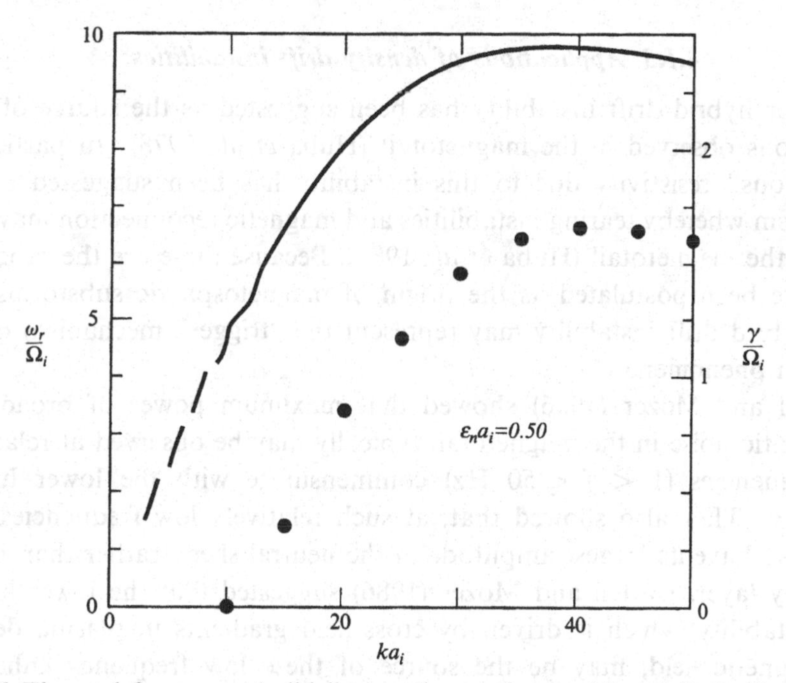

Fig. 4.6 The real frequency (solid line) and growth rate (dotted line) of the lower hybrid drift instability as functions of wavenumber. Here $T_i = 10T_e, v_{ni} = 0.50v_i$ and $\theta = 90°$.

the v_y-dependent part of the ion distribution function. A representative dispersion curve of the instability in this regime is shown in Figure 4.6.

For $|v_{ni}| < v_i$ and unmagnetized ions, the ions are resonant with the lower hybrid drift instability ($\zeta_i^0 \ll 1$), so that, using $\gamma \ll \omega_r$, there is (Gary and

Sanderson, 1979):

$$\omega_r \simeq -\frac{\exp(-\lambda_e)I_0(\lambda_e)k_y v_{ne}}{1 + T_e/T_i - \exp(-\lambda_e)I_0(\lambda_e)}$$

and

$$\gamma \simeq \left(\frac{\pi}{2}\right)^{1/2} \frac{T_e}{T_i} \frac{\exp(-\lambda_e)I_0(\lambda_e)v_{ne}(\omega_r - k_y v_{ni})}{v_i[1 + T_e/T_i - \exp(-\lambda_e)I_0(\lambda_e)]^2}.$$

The wavenumber at maximum growth typically satisfies

$$k_m a_i \simeq \left(\frac{m_i}{m_e}\right)^{1/2}.$$

The lower hybrid drift instability is much less sensitive than the universal drift instability to nonzero β effects, and is generally stabilized only at $\beta > 1$ (Davidson *et al.*, 1977).

4.1.3 Applications of density drift instabilities

The lower hybrid drift instability has been suggested as the source of field fluctuations observed in the magnetotail (Huba *et al.*, 1978). In particular, "anomalous" resistivity due to this instability has been suggested as the mechanism whereby tearing instabilities and magnetic reconnection may take place in the magnetotail (Huba *et al.*, 1981). Because these are the processes that have been postulated as the origin of magnetospheric substorms, the lower hybrid drift instability may represent the "trigger" mechanism of the substorm phenomena.

Cattell and Mozer (1986) showed that maximum power of broadband electrostatic noise in the magnetotail typically may be observed at relatively low frequencies ($1 < f \lesssim 50$ Hz) commensurate with the lower hybrid frequency. They also showed that, at such relatively low frequencies, this noise may have its largest amplitude in the neutral sheet, rather than in the boundary layer. Cattell and Mozer (1986) suggested that the lower hybrid drift instability, which is driven by cross-field gradients in plasma density and magnetic field, may be the source of these low frequency enhanced fluctuations. However, they did not establish a correlation between the field amplitudes and such cross-field sources of free energy, and the source and wave mode for the low frequency component of broadband electrostatic noise appears to be an open question at this writing.

Gary and Eastman (1979) pointed out that the lower hybrid drift instability is the most likely source of field fluctuations observed between 1 and 100 Hz at the magnetopause. Tsurutani *et al.* (1981); LaBelle and Treumann

(1988); and Tsurutani *et al.* (1989) have examined electric and magnetic field fluctuation spectra at the magnetopause. The general conclusion of these studies seems to be that there are no strong correlations between fluctuation amplitudes and magnetopause or magnetosheath parameters, so that, as a consequence, there has been no clear identification of a particular wave or instability.

Conventional quasilinear theory is usually invoked under the assumption that the wave-particle consequences of an instability are to reduce the associated free energy and eventually stabilize the growing mode. Under this assumption, several authors (Davidson and Krall, 1977; Gary, 1980) have used quasilinear theory to calculate the expected rate of cross-field diffusion due to the lower hybrid drift instability. Other authors have used quasilinear theory to calculate cross-field diffusion coefficients due to various other instabilities, and these results were reviewed and applied to the magnetopause by LaBelle and Treumann (1988). Treumann *et al.* (1991) corrected a numerical error of LaBelle and Treumann (1988) to conclude that the lower hybrid drift instability yields the largest wave-particle diffusion coefficient at the magnetopause; they also concluded that this diffusion coefficient is just large enough to explain the thickness of the boundary layer, a region of magnetosheath-like plasma earthward of the magnetopause.

Electromagnetic hybrid simulations (particle ions, fluid electrons) of a magnetopause model with a unidirectional background magnetic field have been run by Gary and Sgro (1990). The lower hybrid drift instability does grow in this simulation and the average density gradient is somewhat reduced. However, at times long compared to an ion gyroperiod, the fluctuations appear to coalesce to longer wavelengths; this result calls into question the traditional model of cross-field diffusion due to wave-particle scattering by short wavelength fluctuations. Thus, the lower hybrid drift instability seems to be the most likely mode to be present at the magnetopause, but observational and computational evidence for its growth and consequent plasma diffusion is not well established.

4.2 Density drift instabilities in accelerated plasmas

In this section we assume that a plasma is subject to a uniform acceleration, such as the acceleration of gravity, which has the same effect on all charged particles. If this acceleration, which we denote by **g**, acts in the direction antiparallel to ∇n, the ion density gradient drift velocity and the ion acceleration drift velocity are parallel (see Figure 4.7.). Under such conditions, the plasma is subject not only to density drift instabilities but

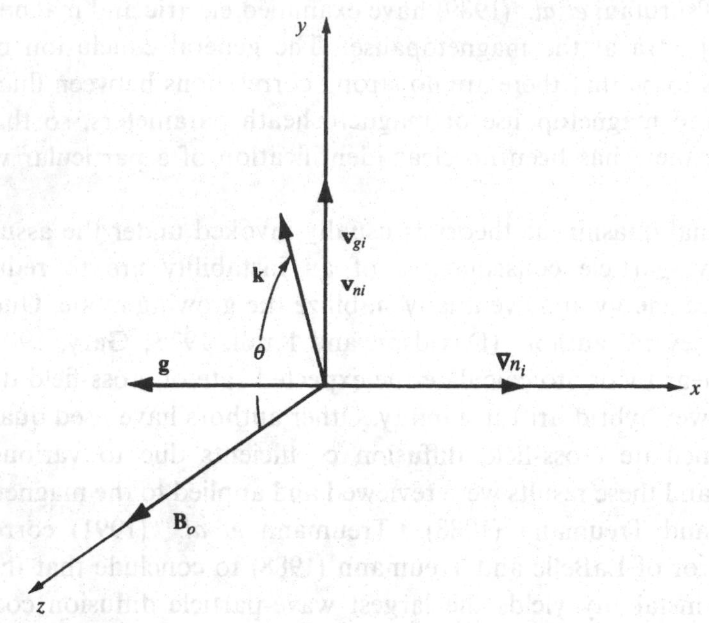

Fig. 4.7 The relationship among the magnetic field, the wavevector, the density gradient, the acceleration, and the ion drift velocities (From Fig. 8 of Sgro *et al.*, 1989).

also to an interchange instability (Spitzer, 1962), the plasma analogue of the Rayleigh-Taylor instability in neutral fluids. This mode is usually treated in the fluid approximation, because it has a maximum growth rate at the long wavelength limit. However, in a collisionless plasma, there is a second, short wavelength ($ka_i \gg 1$) interchange instability which, like its fluid counterpart, has maximum growth at $k_z = 0$.

To derive the full Vlasov dispersion equation appropriate for these instabilities, we follow Gary and Thomsen (1982). An appropriate zeroth-order time-independent distribution function based on constants of the motion (B.7), (B.9) and (B.12) is

$$f_j^{(0)}(x, \mathbf{v}) = \frac{n_j}{(2\pi v_j^2)^{3/2}} \left[1 + \epsilon_n \left(x + \frac{v_y - v_{gj}}{\Omega_j}\right)\right] \exp\left[\frac{-(\mathbf{v} - \mathbf{v}_{gj})^2}{2v_j^2}\right] \quad (4.2.1)$$

where $\mathbf{v}_{gj} = \hat{\mathbf{y}} v_{gj}$ and $v_{gj} \equiv g/\Omega_j$. The zeroth velocity moment of this distribution yields Equation (4.1.1); the first moment implies

$$\mathbf{\Gamma}_j^{(0)}(x) = n_j v_{nj} \hat{\mathbf{y}} + n_j^{(0)}(x) v_{gj} \hat{\mathbf{y}}.$$

Thus both the density gradient and acceleration produce drifts with nonzero currents and so both are sources of free energy available to drive plasma instabilities. Note that, although v_{nj} is independent of species mass, v_{gj} is proportional to m_j, so that it is the ion acceleration drift velocity that is the primary contributor to the free energy here.

Given Equation (4.2.1), the derivation of the linear dispersion equation follows the same procedure as in Section 4.1; the result, which as before is strictly valid only for $\epsilon_n \ll k$ and $\epsilon_n a_j \ll 1$, is

$$K_j(x, \mathbf{k}, \omega) = \frac{k_j^2}{k^2}\left[1 + \frac{\omega - k_y(v_{nj} + v_{gj})}{\sqrt{2}|k_z|v_j}\exp(-\lambda_j)\sum_{m=-\infty}^{\infty} I_m(\lambda_j)Z(\zeta_j^m)\right.$$
$$\left. - \frac{\omega}{\sqrt{2}|k_z|v_j}\frac{\epsilon_n}{k_y}\exp(-\lambda_j)\sum_{m=-\infty}^{\infty} mI_m(\lambda_j)Z(\zeta_j^m)\right] \qquad (4.2.2)$$

where $\lambda_j \equiv k_y^2 a_j^2$ and

$$\zeta_j^m \equiv \frac{\omega - k_y v_{gj} + m\Omega_j}{\sqrt{2}|k_z|v_j}.$$

For those density drift instabilities that satisfy $k^2 \ll k_j^2$, it follows from Equation (4.1.7) that the linear electrostatic dispersion equation is independent of x.

In the limit of perpendicular propagation, $k_z = 0$, and by the asymptotic expansion of the plasma dispersion function,

$$K_j(\mathbf{k}, \omega) = \frac{k_j^2}{k^2}\left\{1 - [\omega - k_y(v_{nj} + v_{gj})]\exp(-\lambda_j)\sum_{m=-\infty}^{\infty}\frac{I_m(\lambda_j)}{\omega - k_y v_{gj} + m\Omega_j}\right.$$
$$\left. + \frac{\omega v_{nj}}{k_y v_j^2}\exp(-\lambda_j)\sum_{m=-\infty}^{\infty}\frac{m\Omega_j I_m(\lambda_j)}{\omega - k_y v_{gj} + m\Omega_j}\right\}. \qquad (4.2.3)$$

4.2.1 *Interchange instabilities*

If $\mathbf{g}$ and ∇n are oppositely directed, the interchange instability may grow in a collisionless plasma. The maximum growth rate of this instability is at perpendicular propagation ($k_z = 0$); this is the fastest-growing mode as long as the density gradient is sufficiently weak ($v_{ni}/v_i \ll 1$) that the lower hybrid drift instability is not excited. In the long wavelength limit, this instability has a growth rate (Rosenbluth *et al.*, 1962)

$$\gamma = (\epsilon_n g)^{1/2}. \qquad (4.2.4)$$

Problem 4.2.1. Under the assumption that the $\omega v_{nj}/k_y v_j^2 ...$ term is of order ϵ_n^2 and may be ignored, show that the use of Equation (4.2.3) in the linear

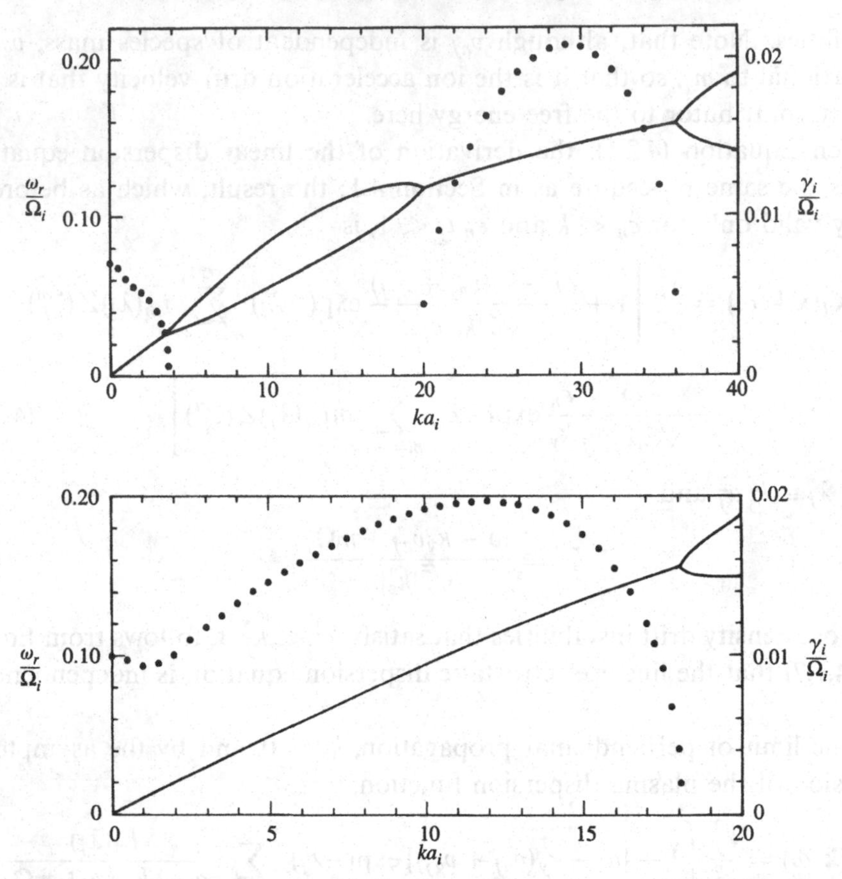

Fig. 4.8 The real frequency (solid lines) and growth rate (dotted lines) of the interchange instabilities as functions of wavenumber. Here $v_{ni} = -0.01v_i$ and $k_z = 0$. In panel (a), $v_{gi} = 0.5v_{ni}$; in panel (*b*), $v_{gi} = v_{ni}$. (From Fig. 1 of Gary and Thomsen, 1982).

dispersion equation (4.1.7) yields equation (4.2.4) in the long wavelength limit, even though (4.2.3) is strictly valid only at $\epsilon_n \ll k$.

Two representative solutions of the full Vlasov dispersion equation are illustrated in Figure 4.8. Panel (a) of this figure represents the situation typical of $|v_{gi}| < |v_{ni}|$ in which there are two distinct regimes for this instability. The long wavelength regime, which we call the "fluid interchange instability," has growth rate (4.2.4) in the long wavelength limit and becomes stable at $k_y a_i$ of order unity. The second, short wavelength regime, which we call the "kinetic interchange instability," emerges at $k_y a_i \gg 1$ with growth rate always larger than that of its fluid counterpart. Panel (b) of Figure 4.8 shows that, in the regime $v_{gi} \simeq v_{ni}$, the two instabilities merge.

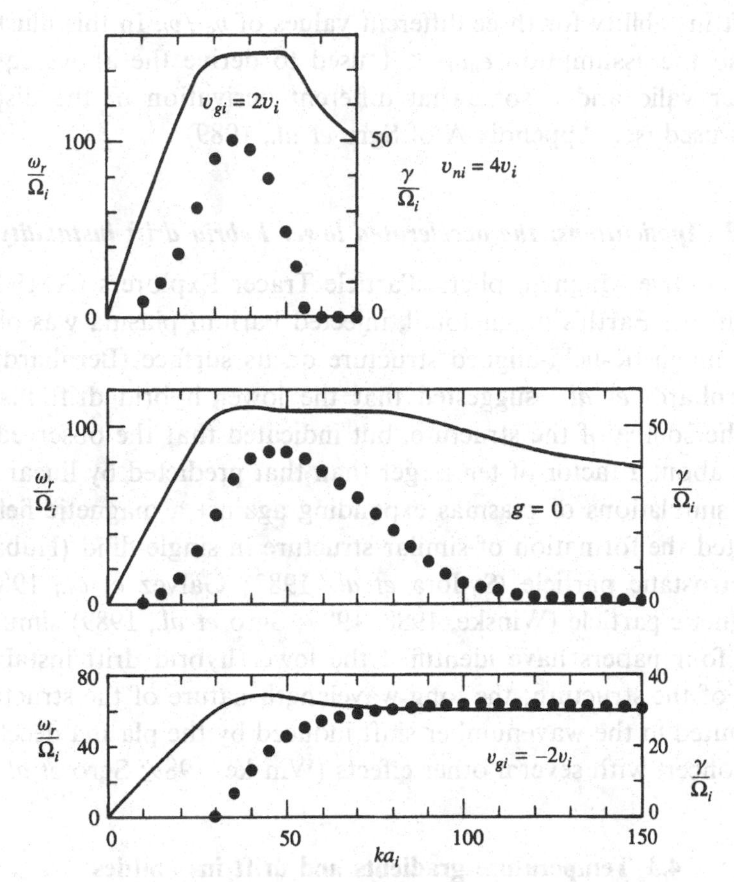

Fig. 4.9 The real frequency (solid lines) and growth rate (dotted lines) of the lower hybrid drift instability as functions of wavenumber for three different values of v_{gi}/v_i. Here $v_{ni} = 4v_i$ and $k_z = 0$ (From Fig. 9 of Sgro *et al.*, 1989).

Detailed properties of the instability in the two regimes are discussed in Gary and Thomsen (1982). Typically the maximum growth rate of the kinetic instability is larger than that of the fluid instability by more than a factor of two, and if $T_e \ll T_i$, γ_m of the kinetic instability follows the same scaling as Equation (4.2.4): $\gamma_m \sim (v_{ni}v_{gi})^{1/2}$.

4.2.2 The lower hybrid drift instability: acceleration effects

Because $k_y a_i \gg 1$ for the kinetic interchange instability, this mode is subsumed into the lower hybrid drift instability as v_{ni}/v_i becomes sufficiently large. If $\mathbf{v}_{gi} \cdot \mathbf{v}_{ni} > 0$, the acceleration enhances γ_m and reduces $k_m a_i$ of the latter instability (Okada *et al.*, 1979; Gary and Thomsen, 1982). This result is also shown in Figure 4.9, which plots the dispersion curves of the lower hybrid drift instability for three different values of v_{gi}/v_i. In this illustration, $v_{ni} = 4v_i$, so the assumption $\epsilon_n a_i \ll 1$ used to derive the above equations is no longer valid and a somewhat different derivation of the dispersion equation is used (see Appendix A of Sgro *et al.*, 1989).

4.2.3 Applications: the accelerated lower hybrid drift instability

During the Active Magnetospheric Particle Tracer Explorers (AMPTE) experiments in the Earth's magnetotail, injected barium plasma was observed to develop magnetic-field-aligned structure on its surface (Bernhardt *et al.*, 1987). Bernhardt *et al.* suggested that the lower hybrid drift instability might be the source of the structure, but indicated that the observed wavelength was about a factor of ten larger than that predicted by linear theory. Computer simulations of plasmas expanding against a magnetic field have demonstrated the formation of similar structure in single-fluid (Huba *et al.*, 1987), electrostatic particle (Sydora *et al.*, 1983; Galvez *et al.*, 1988) and electromagnetic particle (Winske, 1988, 1989; Sgro *et al.*, 1989) simulations. The latter four papers have identified the lower hybrid drift instability as the source of the structure; the long wavelength nature of the structure has been attributed to the wavenumber shift induced by the plasma deceleration acting in concert with several other effects (Winske, 1989; Sgro *et al.*, 1989).

4.3 Temperature gradients and drift instabilties

A temperature gradient perpendicular to $\mathbf{B}_o$ is, like a density gradient, a potential source of free energy. We denote the inverse scale length of the jth species temperature by ϵ_{Tj}. Even in the absence of ∇n, temperature gradients alone can drive several different drift instabilities (Gary and Abraham-Shrauner, 1981). An example of such a mode is the ion temperature drift instability that is driven by ∇T_i and propagates at $\omega_r \ll \Omega_i$, $ka_i \lesssim 1$ and $0 < k_z \ll k_y$. Gary and Abraham-Shrauner (1981) describe the linear properties of this mode and compare it with the universal drift instability. In general, both this mode and the electron temperature drift instability are important only when ∇T_j is the primary source of free energy (i.e. $|\nabla T_j|/T_j \gtrsim |\nabla n|/n$).

Of course, $\epsilon_{Tj} \neq 0$ also can affect the linear dispersion properties of the universal and lower hybrid density drift instabilities (Gary and Sanderson, 1979). Figure 4.10 illustrates the effect of an ion temperature gradient on the lower hybrid drift instability; it should be compared with Figure 4.6, which corresponds to the same parameters except that $\epsilon_{Ti} = 0$. Note that a ∇T_i

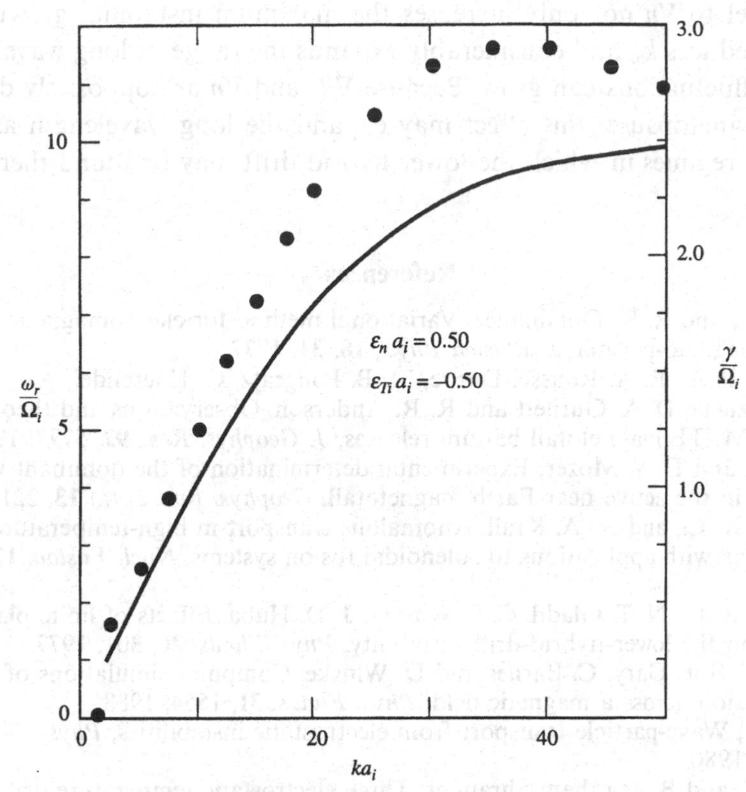

Fig. 4.10 The real frequency (solid line) and growth rate (dotted line) of the lower hybrid drift instability as functions of wavenumber. The parameters are the same as in Figure 4.6 except that $\epsilon_{Ti}a_i = -0.50$.

Table 4.1. *Some electrostatic drift instabilities*

Name	Driven by	Damping	Frequency at maximum growth rate	Wavevector	
Universal drift	v_{ne}	Ion Landau	$\omega_r \sim k_y v_{ne}$	$ka_i < 1$	$0 < k_z << k_y$
Lower hybrid drift	v_{ni}	Electron Landau	$\Omega_i < \omega_r$	$k^2 a_e^2 \sim \frac{T_e}{T_i}$	$k_z = 0$
Kinetic interchange	$v_{gi} + v_{ni}$	Electron Landau	$\omega_r \sim k_y v_{gi}$	$ka_i >> 1$	$k_z = 0$
Ion temperature drift	v_{Ti}	Ion Landau	$\omega_r << \Omega_i$	$ka_i < 1$	$0 < k_z << k_y$

antiparallel to ∇n not only increases the maximum instability growth rate, but also reduces k_m and considerably expands the range of long wavelengths at which fluctuations can grow. Because ∇T and ∇n are oppositely directed at the magnetopause, this effect may expand the long wavelength and low frequency regimes in which the lower hybrid drift may be found there.

References

Berk, H. L., and R. R. Dominguez, Variational method for electromagnetic waves in a magneto-plasma, *J. Plasma Phys.*, **18**, 31, 1977.

Bernhardt, P. A., R. A. Roussel-Dupre, M. B. Pongratz, G. Haerendel, A. Valenzuela, D. A. Gurnett and R. R. Anderson, Observations and theory of the AMPTE magnetotail barium releases, *J. Geophys. Res.*, **92**, 5777, 1987.

Cattell, C., and F. W. Mozer, Experimental determination of the dominant wave mode in the active near-Earth magnetotail, *Geophys. Res. Lett.*, **13**, 221, 1986.

Davidson, R. C., and N. A. Krall, Anomalous transport in high-temperature plasmas with applications to solenoidal fusion systems, *Nucl. Fusion*, **17**, 1313, 1977.

Davidson, R. C., N. T. Gladd, C. S. Wu and J. D. Huba, Effects of finite plasma beta on the lower-hybrid-drift instability, *Phys. Fluids*, **20**, 301, 1977.

Galvez, M., S. P. Gary, C. Barnes and D. Winske, Computer simulations of plasma expansion across a magnetic field, *Phys. Fluids*, **31**, 1554, 1988.

Gary, S. P., Wave-particle transport from electrostatic instabilities, *Phys. Fluids*, **23**, 1193, 1980.

Gary, S. P., and B. Abraham-Shrauner, Three electrostatic temperature drift instabilities, *J. Plasma Phys.*, **25**, 145, 1981.

Gary, S. P., and T. E. Eastman, The lower hybrid drift instability at the magnetopause, *J. Geophys. Res.*, **84**, 7378, 1979.

Gary, S. P., and J. J. Sanderson, Density gradient drift instabilities: oblique propagation at zero beta, *Phys. Fluids*, **21**, 1181, 1978; Erratum, ibid., **22**, 388, 1979.

Gary, S. P., and J. J. Sanderson, Electrostatic temperature gradient drift instabilities, *Phys. Fluids*, **22**, 1500, 1979.

Gary, S. P., and A. G. Sgro, The lower hybrid drift instability at the magnetopause, *Geophys. Res. Lett.*, **17**, 909, 1990.

Gary, S. P., and M. F. Thomsen, Collisionless electrostatic interchange instabilities, *J. Plasma Phys.*, **28**, 551, 1982.

Gary, S. P., P. A. Bernhardt and T. E. Cole, Density drift instabilities and weak collisions, *J. Geophys. Res.*, **88**, 2103, 1983.

Huba, J. D., and S. P. Gary, Finite-β stabilization of the universal drift instability: revisited, *Phys. Fluids*, **25**, 1821, 1982.

Huba, J. D., N. T. Gladd and K. Papadopoulos, Lower-hybrid-drift wave turbulence in the distant magnetotail, *J. Geophys. Res.*, **83**, 5217, 1978.

Huba, J. D., N. T. Gladd and J. F. Drake, On the role of the lower hybrid drift instability in substorm dynamics, *J. Geophys. Res.*, **86**, 5881, 1981.

Huba, J. D., J. G. Lyon and A. B. Hassam, Theory and simulation of the Rayleigh-Taylor instability in the limit of large Larmor radius, *Phys. Rev. Lett.*, **59**, 2971, 1987.

Krall, N. A., and A. W. Trivelpiece, *Principles of Plasma Physics*, McGraw-Hill, New York, 1973.

LaBelle, J., and R. A. Treumann, Plasma waves at the dayside magnetopause, *Space Sci. Revs.*, **47**, 175, 1988.

Okada, S., K. Sato and T. Sekiguchi, Possibility of lower-hybrid-drift instability in laser produced plasma in a uniform magnetic field, *J. Phys. Soc. Japan*, **46**, 355, 1979.

Rosenbluth, M. N., N. A. Krall and N. Rostoker, Finite Larmor radius stabilization of "weakly" unstable confined plasmas, *Nucl. Fusion, Supplement*, 143, 1962.

Sgro, A. G., S. P. Gary and D. S. Lemons, Expanding plasma structure and its evolution toward long wavelengths, *Phys. Fluids*, **B1**, 1890, 1989.

Spitzer, L., *Physics of Fully Ionized Gases* (Second Revised Edition), Interscience, New York, 1962.

Sydora, R. D., J. S. Wagner, L. C. Lee, E. M. Wescott and T. Tajima, Electrostatic Kelvin-Helmholtz instability in a radially injected plasma cloud, *Phys. Fluids*, **26**, 2986, 1983.

Thomsen, M. F., and S. P. Gary, A nonlocal theory of an electrostatic sinusoidal density drift instability, *J. Geophys. Res.*, **87**, 3551, 1982.

Treumann, R. A., J. LaBelle and R. Pottelette, Plasma diffusion at the magnetopause: the case of lower hybrid drift waves, *J. Geophys. Res.*, **96**, 16009, 1991.

Tsurutani, B. T., E. J. Smith, R. M. Thorne, R. R. Anderson, D. A. Gurnett, G. K. Parks, C. S. Lin and C. T. Russell, Wave-particle interactions at the magnetopause: Contributions to the dayside aurora, *Geophys. Res. Lett.*, **8**, 183, 1981.

Tsurutani, B. T., A. L. Brinca, E. J. Smith, R. T. Okida, R. R. Anderson, and T. E. Eastman, A statistical study of ELF-VLF plasma waves at the magnetopause, *J. Geophys. Res.*, **94**, 1270, 1989.

Winske, D., Short-wavelength modes on expanding plasma clouds, *J. Geophys. Res.*, **93**, 2539, 1988.

Winske, D., Development of flute modes on expanding plasma clouds, *Phys. Fluids*, **B1**, 1900, 1989.

5

Electromagnetic fluctuations in uniform plasmas

This chapter begins our study of electromagnetic waves and instabilities in homogeneous plasmas. As before, we consider a collisionless plasma, so that the evolution of the distribution functions is described by the Vlasov equation (1.3.2). In contrast to the previous three chapters, however, we consider a plasma with $\beta > 0$ so that the waves can have fluctuating magnetic as well as electric fields and the full set of Maxwell's equations must be utilized [(1.2.4) through (1.2.7)]. Here and in the following chapters we again assume there exists a steady, uniform, zeroth-order magnetic field $\mathbf{B}_o = \hat{\mathbf{z}}B_o$.

In Section 5.1, we develop a linear theory of electromagnetic fluctuations applicable to a general zeroth-order distribution function $f_j^{(0)}$. Section 5.2 defines several quantities which can be used to identify such fluctuations. In the following chapters, we address specific forms of the distribution functions and their consequent normal modes.

5.1 The electromagnetic dispersion equation

5.1.1 $\mathbf{k} \times \mathbf{B}_o \neq 0$

For propagation at arbitrary angles with respect to $\mathbf{B}_o$ it is convenient to define the dimensionless conductivity of the jth species or component $\mathbf{S}_j(\mathbf{k}, \omega)$:

$$\mathbf{\Gamma}_j^{(1)}(\mathbf{k}, \omega) = -\frac{ik^2c^2}{4\pi e_j \omega}\mathbf{S}_j(\mathbf{k}, \omega) \cdot \mathbf{E}^{(1)}(\mathbf{k}, \omega). \tag{5.1.1}$$

If we combine Equation (5.1.1) with the Maxwell equations (1.2.5) and (1.2.6) we obtain

$$\mathbf{D}(\mathbf{k}, \omega) \cdot \mathbf{E}^{(1)}(\mathbf{k}, \omega) = 0 \tag{5.1.2}$$

where, with $\mathbf{k} = \hat{\mathbf{y}}k_y + \hat{\mathbf{z}}k_z$,

$$
\begin{aligned}
D_{xx} &= \omega^2 - k^2c^2 + k^2c^2 \sum_j S_{xxj} \\
D_{xy} &= k^2c^2 \sum_j S_{xyj} \\
D_{xz} &= k^2c^2 \sum_j S_{xzj} \\
D_{yx} &= k^2c^2 \sum_j S_{yxj} \\
D_{yy} &= \omega^2 - k_z^2c^2 + k^2c^2 \sum_j S_{yyj} \\
D_{yz} &= k_yk_zc^2 + k^2c^2 \sum_j S_{yzj} \\
D_{zx} &= k^2c^2 \sum_j S_{zxj} \\
D_{zy} &= k_yk_zc^2 + k^2c^2 \sum_j S_{zyj} \\
D_{zz} &= \omega^2 - k_y^2c^2 + k^2c^2 \sum_j S_{zzj}.
\end{aligned}
\tag{5.1.3}
$$

The dispersion equation for the various plasma fluctuations of the system is then

$$\det|\mathbf{D}(\mathbf{k}, \omega)| = 0. \tag{5.1.4}$$

This equation is a transcendental equation for the complex variable ω given the real wavevector $\mathbf{k}$; for each choice of plasma parameters, this equation has many solutions $\omega = \omega(\mathbf{k})$. Although, with various approximations, one may obtain analytic expressions for the least damped or the fastest growing of these solutions, our emphasis in the following chapters is on exact numerical solutions of Equation (5.1.4).

From Equation (5.1.2) we may obtain the relative values of the fluctuating field components. In particular,

$$\frac{E_z^{(1)}}{E_y^{(1)}} = \frac{D_{xx}D_{zy} - D_{zx}D_{xy}}{D_{zx}D_{xz} - D_{xx}D_{zz}} \tag{5.1.5}$$

and

$$\frac{E_x^{(1)}}{E_y^{(1)}} = \frac{D_{zz}D_{xy} - D_{xz}D_{zy}}{D_{zx}D_{xz} - D_{xx}D_{zz}}. \tag{5.1.6}$$

To evaluate $\mathbf{S}_j(\mathbf{k}, \omega)$, we use the linear Vlasov equation for electromagnetic waves in a magnetized plasma:

$$\frac{\partial f_j^{(1)}(\mathbf{x},\mathbf{v},t)}{\partial t} + \mathbf{v}\cdot\frac{\partial f_j^{(1)}}{\partial \mathbf{x}} + \frac{e_j}{m_j}\left(\frac{\mathbf{v}\times\mathbf{B}_o}{c}\right)\cdot\frac{\partial f_j^{(1)}}{\partial \mathbf{v}} =$$

$$-\frac{e_j}{m_j}[\mathbf{E}^{(1)}(\mathbf{x},t) + \frac{\mathbf{v}\times\mathbf{B}^{(1)}(\mathbf{x},t)}{c}]\cdot\frac{\partial f_j^{(0)}(\mathbf{v})}{\partial \mathbf{v}}. \tag{5.1.7}$$

From the linear Vlasov equation (5.1.7) and Faraday's equation (1.2.5) the first-order distribution function of the jth species or component may then be written as

$$f_j^{(1)}(\mathbf{k},\mathbf{v},\omega) = -\frac{e_j}{m_j}\int_{-\infty}^{0} d\tau \left[\frac{\partial f_j^{(0)}}{\partial \mathbf{v}'} + \frac{\mathbf{k}}{\omega}\times\left(\mathbf{v}'\times\frac{\partial f_j^{(0)}}{\partial \mathbf{v}'}\right)\right]\cdot$$

$$\mathbf{E}^{(1)}(\mathbf{k},\omega)\ \exp\left[ib_j(\tau,\omega)\right] \tag{5.1.8}$$

where the primed variables denote the unperturbed orbit of a charged particle in $\mathbf{B}_o$ (Appendix B) and $b_j(\tau,\omega)$ is given by Equation (2.3.8).

From this point on, the analytic development depends upon the specific form of the zeroth-order distribution function. The next three chapters treat three specific distributions; here we continue the development of a formalism appropriate for general zeroth-order distributions in a homogeneous plasma.

5.1.2 $\mathbf{k}\times\mathbf{B}_o = 0$

If $\mathbf{k}\times\mathbf{B}_o = 0$ in a homogeneous plasma, it is usually true that $D_{zx} = D_{xz} = 0$ and $D_{zy} = D_{yz} = 0$, so that Equation (5.1.2) factors into two parts, corresponding to (strictly) electrostatic and strictly electromagnetic waves. For electrostatic fluctuations with $E_z^{(1)} \neq 0$,

$$D_{zz} = 0, \tag{5.1.9}$$

which is equivalent to the electrostatic dispersion equation in an unmagnetized plasma, Equation (2.2.9). Thus, all the results of Chapters 2 and 3 for unmagnetized plasmas are applicable to electrostatic waves and instabilities propagating at $\mathbf{k}\times\mathbf{B}_o = 0$.

Problem 5.1.1. At propagation parallel to $\mathbf{B}_o$, prove that Equation (5.1.9) is equivalent to Equation (2.2.9).

In addition, transverse fluctuations with $\mathbf{E}^{(1)}$ and $\mathbf{B}^{(1)}$ perpendicular to $\mathbf{B}_o$

may also propagate along the magnetic field. From the remainder of the dispersion equation one obtains

$$\begin{pmatrix} D_{xx} & D_{xy} \\ D_{yx} & D_{yy} \end{pmatrix} \begin{pmatrix} E_x^{(1)} \\ E_y^{(1)} \end{pmatrix} = 0 \qquad (5.1.10)$$

from which follows the strictly electromagnetic dispersion equation for $\mathbf{k} \times \mathbf{B}_o = 0$:

$$\omega^2 - k^2c^2 + k^2c^2 \sum_j S_j^{\mp}(\mathbf{k}, \omega) = 0. \qquad (5.1.11)$$

To evaluate the dimensionless conductivity for a general zeroth-order distribution function, we integrate the linear Vlasov equation by the method of unperturbed orbits. We obtain

$$f_j^{(1)}(\mathbf{k}, \mathbf{v}, \omega) =$$

$$-\frac{e_j}{m_j\omega} \int_{-\infty}^{0} d\tau\, \mathbf{v}' \cdot \mathbf{E}^{(1)} \left[k_z \frac{\partial f_j^{(0)}}{\partial v_z} + \frac{(\omega - k_z v_z)}{v_\perp} \frac{\partial f_j^{(0)}}{\partial v_\perp} \right] \exp\left[i(k_z v_z - \omega)\tau\right].$$

At $\mathbf{k} \times \mathbf{B}_o = 0$, electromagnetic fluctuations are transverse to $\mathbf{B}_o$ and circularly polarized, so it is convenient to work in terms of the unit vectors

$$\hat{\mathbf{e}}_\pm = \frac{\hat{\mathbf{x}} \pm i\hat{\mathbf{y}}}{\sqrt{2}}$$

so that the perpendicular component of any vector may be written

$$\mathbf{a}_\perp = a_+\hat{\mathbf{e}}_- + a_-\hat{\mathbf{e}}_+$$

where

$$a_\pm = \frac{a_x \pm ia_y}{\sqrt{2}}. \qquad (5.1.12)$$

From Equations (B.1) and (B.2) it follows that

$$v_\pm(t') = v_\pm(t)\exp[\mp i\Omega_j\tau]$$

so that evaluation of the integral yields

$$f_j^{(1)}(\mathbf{k}, \mathbf{v}, \omega) = \sum f_{j\mp}^{(1)} = \frac{ie_j}{m_j\omega} \left[k_z \frac{\partial f_j^{(0)}}{\partial v_z} + \frac{(\omega - k_z v_z)}{v_\perp} \frac{\partial f_j^{(0)}}{\partial v_\perp} \right]$$

$$\sum \frac{v_\pm E_\mp^{(1)}}{k_z v_z - \omega \mp \Omega_j} \qquad (5.1.13)$$

where $\sum$ indicates summation over the $\pm$ signs.

Using Equation (2.2.2) in Maxwell's equations (1.2.5) and (1.2.6) and the assumption that $\mathbf{k} \cdot \mathbf{E}^{(1)} = 0$, we obtain

$$\mathbf{E}^{(1)}(\mathbf{k}, \omega)\,(\omega^2 - k^2c^2) = -4\pi\omega i \mathbf{J}^{(1)}(\mathbf{k}, \omega)$$

so that if one defines $S_j^{\pm}$ by

$$\Gamma_{j\mp}^{(1)}(\mathbf{k}, \omega) = -\frac{ik^2c^2}{4\pi e_j \omega} S_j^{\pm}(\mathbf{k}, \omega) \mathbf{E}_{\mp}^{(1)}(\mathbf{k}, \omega)$$

the linear dispersion equation becomes Equation (5.1.11) where

$$S_j^{\pm}(\mathbf{k}, \omega) = -\frac{\omega_j^2}{2k^2c^2 n_j} \int d^3v\, v_\perp \frac{k_z v_\perp \frac{\partial f_j^{(0)}}{\partial v_z} + (\omega - k_z v_z) \frac{\partial f_j^{(0)}}{\partial v_\perp}}{k_z v_z - \omega \mp \Omega_j}. \tag{5.1.14}$$

In the special case that the zeroth-order distribution function is isotropic, $f_j^{(0)}$ is a function only of $v^2 = v_z^2 + v_\perp^2$, and the dimensionless conductivity assumes a simpler form

$$S_j^{\pm}(\mathbf{k}, \omega) = \frac{\omega_j^2 \omega}{k^2c^2 n_j} \int d^3v\, \frac{f_j^{(0)}(v)}{k_z v_z - \omega \mp \Omega_j}. \tag{5.1.15}$$

Problem 5.1.2. Derive Equation (5.1.15) and use it to show that, for a plasma in which all species are isotropic, electromagnetic fluctuations at parallel propagation are nongrowing (Brinca, 1990).

If one assumes $|\gamma| \ll |\omega_r|$ and uses Equation (2.2.10), then the real and imaginary parts of the linear dispersion equation yield estimates for ω_r and γ, respectively. In particular, if the zeroth-order distribution function is separable such that $f_j^{(0)}(v_z, v_\perp^2) = f_j^{(0)}(v_z)\, F_j^{(0)}(v_\perp^2)$ with

$$\int F_j^{(0)}(v_\perp^2) dv_x dv_y = 1$$

and

$$\int v_\perp^2 F_j^{(0)}(v_\perp^2) dv_x dv_y = \langle v_\perp^2 \rangle$$

then it follows that for $k_z > 0$

$$\gamma \simeq \frac{\pi}{2\omega_r} \sum_j \frac{\omega_j^2}{n_j} \left[\frac{\langle v_\perp^2 \rangle}{2} \frac{\partial f_j^{(0)}(v_z)}{\partial v_z} \pm \frac{\Omega_j}{k_z} f_j^{(0)}(v_z) \right]_{v_z = (\omega_r \pm \Omega_j)/k_z} \tag{5.1.16}$$

Note that there are two important differences between this equation and the corresponding expression for the damping/growth rate of electrostatic fluctuations, Equation (2.2.11). First, the cyclotron resonance of the electromagnetic waves dictate that the reduced distribution function is evaluated at the cyclotron resonance speed, rather than at the phase speed of the wave. Second, although the terms on the right-hand side of Equation (2.2.11) are all proportional to the slope of the reduced distribution functions, this is not the case in Equation (5.1.16).

Problem 5.1.3. Verify Equation (5.1.16) and derive the approximate damping rate for electromagnetic waves at parallel propagation in a plasma with Maxwellian components (Equation (6.1.8)).

5.2 Identifiers of electromagnetic fluctuations

5.2.1 Polarization

An important property of electromagnetic waves and instabilities is their polarization. Throughout this book, we follow the plasma physics definition of polarization (Stix, 1962), which defines right- or left-handed to be the sense of rotation in time at a fixed point in space of a fluctuating field vector when viewed in the direction parallel to the magnetic field at positive real frequency. Under this definition, a right-hand mode propagating either parallel or anti-parallel to $\mathbf{B}_o$ possesses fluctuating field vectors that rotate in the same sense as the gyromotion of an electron with $v_z = 0$, and corresponds to the magnetosonic/whistler wave in a stable plasma at parallel propagation. Similarly, a left-hand mode rotates in the same sense as a gyrating ion and corresponds to the Alfvén/ion cyclotron wave in a stable plasma at parallel propagation.

If the definition of Stix' (1962) Equation (1.31) is generalized to include $\omega_r < 0$, the polarization P with $\mathbf{k} = \hat{\mathbf{y}}k_y + \hat{\mathbf{z}}k_z$ is

$$P = \frac{E_y^{(1)}}{iE_x^{(1)}} \frac{\omega_r}{|\omega_r|}$$

where $E_\alpha^{(1)}$ is the first-order fluctuating electric field component in the α direction. The polarization of an electromagnetic fluctuation at parallel or antiparallel propagation may be written as

$$P \equiv \pm \frac{\omega_r}{|\omega_r|} \tag{5.2.1}$$

where $\pm$ refers to the corresponding signs in Equation (5.1.11). In this case, $P = +1$ corresponds to a right-hand circularly polarized mode and $P = -1$ to a left-hand circular mode. In the more general case of oblique propagation, polarization is elliptic and $P > 0$ corresponds to right-hand, $P < 0$ to left-hand waves.

5.2.2 Correlation functions

Correlation functions are the ensemble-averaged scalar products of fluctuating quantities in a medium. Self-correlations of the vector fluctuation $\delta\mathbf{f}$ are of the form

$$C_{\mathbf{ff}}(\mathbf{x}, \mathbf{x}', t, t') \equiv \langle \delta\mathbf{f}(\mathbf{x}, t) \cdot \delta\mathbf{f}(\mathbf{x}', t') \rangle \tag{5.2.2}$$

whereas cross-correlations are usually written in the symmetric form

$$C_{\mathbf{fg}}(\mathbf{x}, \mathbf{x}', t, t') \equiv \frac{1}{2}\langle \delta\mathbf{f}(\mathbf{x}, t) \cdot \delta\mathbf{g}(\mathbf{x}', t') + \delta\mathbf{g}(\mathbf{x}, t) \cdot \delta\mathbf{f}(\mathbf{x}', t') \rangle. \tag{5.2.3}$$

In a plasma in which the ensemble averaged quantities are spatially homogeneous, correlation functions are functions only of $\mathbf{x} - \mathbf{x}' = \mathbf{r}$ (e.g. Montgomery, 1971). Similarly, if the ensemble is stationary, correlations are functions only of $t - t' = \tau$. Under these conditions, one may take the Fourier transforms of this function to obtain

$$\begin{aligned} \langle \delta\mathbf{f} \cdot \delta\mathbf{g} \rangle_{\mathbf{k}\omega} &\equiv \int d^3r \, d\tau \, \exp[-i\mathbf{k} \cdot (\mathbf{x} - \mathbf{x}') + i\omega(t - t')] C_{\mathbf{fg}}(\mathbf{x} - \mathbf{x}', t - t') \quad (5.2.4) \\ &= \frac{1}{2}[\delta\mathbf{f}^*(\mathbf{k}, \omega) \cdot \delta\mathbf{g}(\mathbf{k}, \omega) + \delta\mathbf{g}^*(\mathbf{k}, \omega) \cdot \delta\mathbf{f}(\mathbf{k}, \omega)] \\ &= \mathrm{Re}[\delta\mathbf{f}^*(\mathbf{k}, \omega) \cdot \delta\mathbf{g}(\mathbf{k}, \omega)]. \end{aligned}$$

The fundamental self-correlation functions of low frequency fluctuations in a plasma are the magnetic energy density $\langle \delta\mathbf{B} \cdot \delta\mathbf{B} \rangle / 8\pi$ and the jth species density $\langle \delta n_j \delta n_j \rangle$ and velocity $\langle \delta\mathbf{v}_j \cdot \delta\mathbf{v}_j \rangle$. At frequencies well above the proton cyclotron frequency, the electric field energy density $\langle \delta\mathbf{E} \cdot \delta\mathbf{E} \rangle / 8\pi$ usually becomes larger, and a more important indicator of fluctuation properties, than the magnetic energy density. Cross-correlations of interest include the magnetic helicity $\langle \delta\mathbf{A} \cdot \delta\mathbf{B} \rangle$ where $\mathbf{A}$ is the magnetic vector potential that defines $\mathbf{B}$ via $\mathbf{B} = \nabla \times \mathbf{A}$, the jth species cross-helicity $\langle \delta\mathbf{v}_j \cdot \delta\mathbf{b} \rangle$, where $\delta\mathbf{b} \equiv \delta\mathbf{B}/(4\pi n_p m_p)^{1/2}$, and the parallel compressibility, $\langle \delta n_j \delta B_{\parallel} \rangle$.

If one assumes a collisionless plasma with only damped modes, there is an elegant theory that predicts the electric and magnetic field self-correlation functions (e.g. Akhiezer *et al.*, 1966; Montgomery, 1971). However, this theory breaks down as the normal modes of the plasma become unstable. Thus, it has not proven useful in predicting the properties of the nonthermal enhanced fluctuations that result from instability growth and that often are the basis of wave-particle transport.

5.2.3 Transport ratios

Although electric and magnetic field self-correlations become singular as a plasma mode makes the transition from damping to growth, appropriately constructed ratios of correlation functions are well behaved through this transition. To understand this, consider Equation (5.1.1), which states that, in Fourier transformed variables, the fluctuating particle flux of the jth species is the product of a tensor conductivity and a fluctuating field. Equations (5.1.5) and (5.1.6) demonstrate that $E_x^{(1)}(\mathbf{k},\omega)$ and $E_z^{(1)}(\mathbf{k},\omega)$ are each proportional to $E_y^{(1)}(\mathbf{k},\omega)$. Thus, if we divide one of the correlation functions defined in the above section by a field self-correlation function, the result (which we term a "transport ratio") is independent of the field self-correlation and, if fluctuating densities or velocities are involved, depends primarily on the dimensionless conductivity which is a well-behaved function of $\mathbf{k}$ and ω.

In our study of electromagnetic waves and instabilities, we will evaluate six different transport ratios. Two of these are explicit functions only of fluctuating fields and potentials, whereas the other four are functions of both fluctuating fields and plasma quantities. In Fourier transformed variables they are the electric/magnetic field energy ratio

$$\sigma_{\mathbf{EE}}(\mathbf{k},\omega) \equiv \frac{\langle \delta\mathbf{E}\cdot\delta\mathbf{E}\rangle_{\mathbf{k}\omega}}{\langle \delta\mathbf{B}\cdot\delta\mathbf{B}\rangle_{\mathbf{k}\omega}}, \tag{5.2.5}$$

the dimensionless helicity (Matthaeus and Goldstein, 1982),

$$\sigma(\mathbf{k},\omega) \equiv \frac{k\langle \delta\mathbf{A}\cdot\delta\mathbf{B}\rangle_{\mathbf{k}\omega}}{\langle \delta\mathbf{B}\cdot\delta\mathbf{B}\rangle_{\mathbf{k}\omega}}, \tag{5.2.6}$$

the Alfvén ratio of the jth species (Matthaeus and Goldstein, 1982),

$$R_{Aj}(\mathbf{k},\omega) \equiv \frac{\langle \delta\mathbf{v}_j\cdot\delta\mathbf{v}_j\rangle_{\mathbf{k}\omega}}{\langle \delta\mathbf{b}\cdot\delta\mathbf{b}\rangle_{\mathbf{k}\omega}}, \tag{5.2.7}$$

the dimensionless cross-helicity of the jth species (Matthaeus and Goldstein, 1982),

$$\sigma_{cj}(\mathbf{k},\omega) \equiv \frac{2\langle \delta\mathbf{v}_j\cdot\delta\mathbf{b}\rangle_{\mathbf{k}\omega}}{\langle \delta\mathbf{v}_j\cdot\delta\mathbf{v}_j\rangle_{\mathbf{k}\omega}+\langle \delta\mathbf{b}\cdot\delta\mathbf{b}\rangle_{\mathbf{k}\omega}}, \tag{5.2.8}$$

the compressibility of the jth species (Gary, 1986),

$$C_j(\mathbf{k},\omega) \equiv \frac{\langle \delta n_j\delta n_j\rangle_{\mathbf{k}\omega}}{n_j^2}\frac{B_o^2}{\langle \delta\mathbf{B}\cdot\delta\mathbf{B}\rangle_{\mathbf{k}\omega}}, \tag{5.2.9}$$

and the dimensionless cross-correlation between the fluctuating density of the jth component and the fluctuating magnetic field component parallel to the average magnetic field $\delta B_\parallel$ (Lacombe *et al.*, 1990),

$$C_{\parallel j}(\mathbf{k},\omega) \equiv \frac{\langle \delta n_j \delta B_\parallel \rangle_{\mathbf{k}\omega}}{n_j B_o} \frac{B_o^2}{\langle \delta B_\parallel \delta B_\parallel \rangle_{\mathbf{k}\omega}} \tag{5.2.10}$$

which we term the "parallel compressibility" to distinguish it from $C_j(\mathbf{k},\omega)$.

The transport ratios evaluated here are constructed by replacing the Fourier/Fourier fluctuating quantities of arbitrary order (denoted by a preceding δ in Equations (5.2.5) through (5.2.10)) with the Fourier/Laplace transforms of fluctuating quantities from linear theory (denoted by a superscript (1)) obtained from Equations (5.1.1), the equation of continuity, and Maxwell's equations. These functions of wavevector and frequency are reduced to functions of wavevector alone by the substitution $\omega = \omega(\mathbf{k})$ where $\omega(\mathbf{k})$ is the complex frequency that satisfies the linear dispersion equation, (5.1.4).

In the evaluation of several of the transport coefficients it is convenient to use $\mathbf{E}_T^{(1)} \equiv \mathbf{E}^{(1)} - \hat{\mathbf{k}}(\hat{\mathbf{k}}\cdot\mathbf{E}^{(1)})$, the fluctuating electric field component transverse to $\mathbf{k}$, which may be obtained from Faraday's equation (1.2.5):

$$\frac{|\mathbf{E}_T^{(1)}|^2}{|\mathbf{B}^{(1)}|^2} = \frac{|\omega|^2}{k^2c^2}.$$

Then it follows that

$$\sigma_{\mathbf{EE}} = \frac{|\omega|^2}{k^2c^2}\frac{|\mathbf{E}^{(1)}|^2}{|\mathbf{E}_T^{(1)}|^2} \tag{5.2.11}$$

where

$$\frac{|\mathbf{E}_T^{(1)}|^2}{|\mathbf{E}^{(1)}|^2} = 1 - \frac{|k_y E_y^{(1)} + k_z E_z^{(1)}|^2}{k^2(|E_x^{(1)}|^2 + |E_y^{(1)}|^2 + |E_z^{(1)}|^2)}$$

which can be evaluated through the use of Equations (5.1.5) and (5.1.6).

If one imposes the Coulomb gauge ($\nabla \cdot \mathbf{A} = 0$), it then follows from Equation (5.2.6) that

$$\sigma(\mathbf{k}) = \frac{2}{k|E_T^{(1)}|^2}\mathrm{Im}\left(k_y E_x^{(1)} E_z^{(1)^*} - k_z E_x^{(1)} E_y^{(1)^*}\right). \tag{5.2.12}$$

In the limit of $\mathbf{k} \times \mathbf{B}_o = 0$, this reduces to

$$\sigma(\mathbf{k}) = \pm\frac{k_z}{k} \tag{5.2.13}$$

where $\pm$ corresponds to the $\pm$ in the field-aligned dispersion equation

(5.1.11). If $\sigma(\mathbf{k}) = +1$, the wave fields exhibit a left-hand sense of rotation with respect to $\mathbf{k}$, while $\sigma(\mathbf{k}) = -1$ implies a right-handed field structure (Smith *et al.*, 1983). To avoid confusion with polarization terminology, we will use the terms positive and negative helicities, respectively, to describe such fluctuations.

Problem 5.2.1. Show that the + sign on $S_j^{\pm}$ in Equation (5.1.11) corresponds to a positive helicity, i.e. a left-hand sense of spatial rotation with respect to the wavevector. Furthermore, if $\omega_r > 0$ and $B_o > 0$, show that the + sign of $S_j^{\pm}$ corresponds to right-hand polarization, i.e. the temporal sense of rotation of an electron at a fixed point in z. (Hint: consider the fluctuating magnetic field $B_+^{(1)}(z,t)$ expressed in terms of its x- and y- components via Equation (5.1.12).)

Because the equation of continuity implies a direct proportionality between the fluctuating species density and the fluctuating species particle flux in a homogeneous plasma, it follows that the compressibility of the jth species is

$$C_j(\mathbf{k}) = \frac{\Omega_j^2 c^2}{\omega_j^4} \frac{k^2 c^2}{|\omega|^2} \frac{|\mathbf{k} \cdot \mathbf{S}_j \cdot \mathbf{E}^{(1)}|^2}{|\mathbf{E}_T^{(1)}|^2}. \tag{5.2.14}$$

Similarly, the parallel compressibility of the jth species may be evaluated as

$$C_{\parallel j} = -\frac{\Omega_j}{\omega_j^2} \frac{k^2 c^2}{k_y} \mathrm{Im} \left(\frac{\mathbf{k} \cdot \mathbf{S}_j \cdot \mathbf{E}^{(1)}}{\omega E_x^{(1)}} \right). \tag{5.2.15}$$

Both the Alfvén wave and the parallel magnetosonic wave at $\theta = 0°$ have $n_j^{(1)} = 0$, so both C_j and $C_{\parallel j}$ of any species for these modes in this limit are zero. The ion acoustic wave at $\mathbf{k} \times \mathbf{B}_o = 0$ is electrostatic, in which case both compressibilities are undefined.

If the average drift velocity of the jth species is zero, then the fluctuating particle flux density is proportional to the fluctuating species velocity, and the dimensionless cross-helicity may be written as

$$\sigma_{cj}(\mathbf{k}) = \frac{2}{1 + R_{Aj}} \frac{n_p}{n_j} \frac{e_p}{e_j} \frac{c}{\omega_p} \frac{1}{|\mathbf{E}_T^{(1)}|^2} \mathrm{Im}\{\mathbf{k} \cdot [\mathbf{E}^{(1)*} \times (\mathbf{S}_j \cdot \mathbf{E}^{(1)})]\} \tag{5.2.16}$$

In the combined limits of $\theta = 0°$ and $\omega_r \ll \Omega_p$,

$$\sigma_{cj}(\mathbf{k}) = \frac{2}{1 + R_{Aj}} \frac{\omega_r}{k_z v_A} \tag{5.2.17}$$

Problem 5.2.2. Derive Equations (5.2.16) and (5.2.17).

Also in the limit of zero average velocity, the Alfvén ratio of the jth species is

$$R_{Aj}(\mathbf{k}) = \frac{n_p}{n_j}\frac{m_p}{m_j}\frac{k^2c^2}{\omega_j^2}\frac{|\mathbf{S}_j \cdot \mathbf{E}^{(1)}|^2}{|\mathbf{E}_T^{(1)}|^2}$$

Because ideal magnetohydrodynamic theory predicts $\mathbf{v}_p^{(1)} = \pm\mathbf{B}^{(1)}/(4\pi n_p m_p)^{1/2}$ at $\theta = 0°$, $R_{Ap}(\mathbf{k})$ is sometimes used as a measure of the "Alfvén-ness" of low frequency fluctuations. In the combined limits of $\theta = 0°$ and $\omega_r \ll \Omega_p$, $R_{Aj} \simeq |\omega|^2/k^2v_A^2$ for both electrons and ions in electromagnetic fluctuations. The Alfvén ratio is undefined for ion acoustic fluctuations at parallel propagation.

In the following chapters, we use transport ratios, rather than electric or magnetic polarizations, to supplement $\omega(\mathbf{k})$ in demonstrating properties of waves and instabilities at frequencies below the ion cyclotron frequency. We do this because transport ratios constructed from the Fourier/Laplace transforms of symmetric correlation functions are real, whereas polarizations at oblique propagation derived from Vlasov theory are in general complex. Because our goal is to facilitate comparison with spacecraft data, and because such data typically does not provide complex polarizations, we emphasize transport ratios in what follows.

Note that, as is evident from Equations (5.2.13) and (5.2.17), a reversal in the direction of wave propagation leads to a change in sign of both the helicity and the cross-helicity. Thus, although these quantities can provide information about the direction of propagation of waves in an isotropic plasma, their lack of a unique value reduces their utility in identifying waves. The compressibility and the Alfvén ratio are by definition non-negative quantities and obviously do not change sign under a reversal of propagation direction; they therefore rank considerably higher in terms of utility for mode identification. The parallel compressibility also does not reverse sign under a change in propagation direction. However, for predominantly transverse fluctuations, $|\delta B_\parallel|$ is often relatively small so that $C_{\parallel j}$ is observed to have relatively large fluctuations (Lacombe *et al.* 1990), thereby reducing the utility of this quantity as a mode identifier.

References

Akhiezer, A. I., I. A. Akhiezer, R. V. Polovin, A. G. Sitenko, and K. N. Stepanov, *Collective Oscillations in a Plasma*, MIT Press, Cambridge, MA, 1967.

Brinca, A. L., On the electromagnetic stability of isotropic populations, *J. Geophys. Res.*, **95**, 221, 1990.

Gary, S. P., Low-frequency waves in a high-beta collisionless plasma: polarization, compressibility and helicity, *J. Plasma Phys.*, **35**, 431, 1986.

Lacombe, C., E. Kinzelin, C. C. Harvey, D. Hubert, A. Mangeney, J. Elaoufir, D. Burgess, and C. T. Russell, Nature of the turbulence observed by ISEE 1-2 during a quasi-perpendicular crossing of the Earth's bow shock, *Ann. Geophys.*, **8**, 489, 1990.

Matthaeus, W. H., and M. L. Goldstein, Measurement of the rugged invariants of magnetohydrodynamics turbulence in the solar wind, *J. Geophys. Res.*, **87**, 6011, 1982.

Montgomery, D. C., *Theory of the Unmagnetized Plasma*, Gordon and Breach, New York, 1971.

Smith, C. W., M. L. Goldstein, and W. H. Matthaeus, Turbulence analysis of the Jovian upstream 'wave' phenomenon, *J. Geophys. Res.*, **88**, 5581, 1983; Correction, *J. Geophys. Res.*, **89**, 9159, 1984.

Stix, T. H., *The Theory of Plasma Waves*, McGraw-Hill, New York, 1962.

6
Electromagnetic waves in uniform plasmas

In this chapter we consider electromagnetic waves in homogeneous, magnetized, collisionless plasmas with Maxwellian zeroth-order distribution functions (Equation (2.1.1)). As for the electrostatic fluctuations described in Chapter 2, the use of such distributions implies that the solutions of the linear dispersion equation yield only waves, i.e. normal modes of the plasma that do not grow in time. Section 6.1 outlines the derivation of the elements of the dimensionless conductivity tensor, Section 6.2 describes solutions of the dispersion equation for various frequency ranges, and Section 6.3 is a brief summary and discussion of applications.

6.1 The electromagnetic dispersion equation

6.1.1 $\mathbf{k} \times \mathbf{B}_o \neq 0$

For isotropic distribution functions, including the Maxwellian $f_j^{(0)}$ considered in this chapter, the $\mathbf{v} \times \mathbf{B}^{(1)} \cdot \partial f_j^{(0)}/\partial \mathbf{v}$ term on the right-hand side of Equation (5.1.7) vanishes and the linear Vlasov equation reduces to Equation (2.3.1). Then we integrate over the unperturbed orbits, Equations (B.1) through (B.6), to obtain the first-order distribution function, which through the use of the Maxwellian distribution becomes

$$f_j^{(1)}(\mathbf{k}, \mathbf{v}, \omega) = \frac{e_j}{T_j} f_j^{(M)}(v) \int_{-\infty}^{0} d\tau \, \mathbf{v}' \cdot \mathbf{E}^{(1)}(\mathbf{k}, \omega) \exp[ib_j(\tau, \omega)] \qquad (6.1.1)$$

where $b_j(\tau, \omega)$ is given by Equation (2.3.8). Then, via Equation (5.1.1)

$$\mathbf{S}_j^{(M)}(\mathbf{k}, \omega) = \frac{ik_j^2 \omega}{k^2 n_j c^2} \int d^3v \, \mathbf{v} f_j^{(M)}(v) \int_{-\infty}^{0} d\tau \, \mathbf{v}' \exp[ib_j(\tau, \omega)] \qquad (6.1.2)$$

where the superscript (M) denotes a conductivity derived from the Maxwellian distribution function. Evaluating $\mathbf{S}_j^{(M)}$ through the method outlined in Appendix C we obtain

$$
\begin{aligned}
S_{xxj}^{(M)} &= \frac{\omega_j^2}{k^2c^2}\zeta_j^0 \exp(-\lambda_j) \sum_{m=-\infty}^{\infty} \left\{ 2\lambda_j[I_m(\lambda_j) - I_m'(\lambda_j)] + \frac{m^2}{\lambda_j} I_m(\lambda_j) \right\} Z(\zeta_j^m) \\
S_{xyj}^{(M)} &= i\frac{\omega_j^2}{k^2c^2}\zeta_j^0 \exp(-\lambda_j) \sum_{m=-\infty}^{\infty} m[I_m(\lambda_j) - I_m'(\lambda_j)] Z(\zeta_j^m) \\
S_{xzj}^{(M)} &= i\frac{\omega_j^2}{k^2c^2}\frac{|k_z|}{\sqrt{2}k_z}\zeta_j^0 \frac{k_y v_j}{\Omega_j} \exp(-\lambda_j) \sum_{m=-\infty}^{\infty} [I_m(\lambda_j) - I_m'(\lambda_j)] Z'(\zeta_j^m) \\
S_{yxj}^{(M)} &= -S_{xyj}^{(M)} \\
S_{yyj}^{(M)} &= \frac{\omega_j^2}{k^2c^2}\zeta_j^0 \frac{\exp(-\lambda_j)}{\lambda_j} \sum_{m=-\infty}^{\infty} m^2 I_m(\lambda_j) Z(\zeta_j^m) \qquad (6.1.3) \\
S_{yzj}^{(M)} &= \frac{\omega_j^2}{k^2c^2}\frac{|k_z|}{\sqrt{2}k_z}\zeta_j^0 \frac{\Omega_j}{k_y v_j} \exp(-\lambda_j) \sum_{m=-\infty}^{\infty} m I_m(\lambda_j) Z'(\zeta_j^m) \\
S_{zxj}^{(M)} &= -S_{xzj}^{(M)} \\
S_{zyj}^{(M)} &= S_{yzj}^{(M)} \\
S_{zzj}^{(M)} &= -\frac{\omega_j^2}{k^2c^2}\zeta_j^0 \exp(-\lambda_j) \sum_{m=-\infty}^{\infty} I_m(\lambda_j) \zeta_j^m Z'(\zeta_j^m)
\end{aligned}
$$

where $\lambda_j \equiv (k_y a_j)^2$ and

$$\zeta_j^m \equiv \frac{\omega + m\Omega_j}{\sqrt{2}|k_z|v_j}.$$

6.1.2 $\mathbf{k} \times \mathbf{B}_o = 0$

If $\mathbf{k}$ is parallel or antiparallel to $\mathbf{B}_o$, $k_y = 0$. In this case we write $k = |k_z|$ and the above expressions reduce to

$$
\begin{aligned}
S_{xxj}^{(M)} &= \frac{1}{2}\frac{\omega_j^2}{k^2c^2}\zeta_j^0[Z(\zeta_j^{+1}) + Z(\zeta_j^{-1})] \\
S_{xyj}^{(M)} &= -\frac{i}{2}\frac{\omega_j^2}{k^2c^2}\zeta_j^0[Z(\zeta_j^{+1}) - Z(\zeta_j^{-1})] \\
S_{xzj}^{(M)} &= 0 \\
S_{yxj}^{(M)} &= -S_{xyj}^{(M)} \\
S_{yyj}^{(M)} &= S_{xxj}^{(M)} \qquad (6.1.4)
\end{aligned}
$$

$$S_{yzj}^{(M)} = 0$$
$$S_{zxj}^{(M)} = 0$$
$$S_{zyj}^{(M)} = 0$$
$$S_{zzj}^{(M)} = -\frac{k_j^2}{k^2}\frac{v_j^2}{c^2}(\zeta_j^0)^2 Z'(\zeta_j^0).$$

Thus, if $\mathbf{k} \times \mathbf{B}_o = 0$, Equation (5.1.2) factors into two parts, corresponding to (strictly) electrostatic and strictly electromagnetic fluctuations. For electrostatic waves with $E_z^{(1)} \neq 0$,

$$D_{zz} = 0, \tag{6.1.5}$$

which is equivalent to the electrostatic dispersion equation in an unmagnetized plasma, Equation (2.2.8) with Equation (2.2.12). Thus, for propagation parallel or antiparallel to $\mathbf{B}_o$, all of the results of Section 2.1 are appropriate.

In addition, transverse waves with $\mathbf{E}^{(1)}$ and $\mathbf{B}^{(1)}$ perpendicular to $\mathbf{B}_o$ may also propagate along the magnetic field. From the remainder of the dispersion equation one obtains Equation (5.1.10), from which follows the strictly electromagnetic dispersion equation for $\mathbf{k} \times \mathbf{B}_o = 0$:

$$\omega^2 - k^2c^2 + k^2c^2 \sum_j S_j^{\pm}(\mathbf{k}, \omega) = 0 \tag{6.1.6}$$

where

$$S_j^{\pm}(\mathbf{k}, \omega) = \frac{\omega_j^2}{k^2c^2}\zeta_j^0 Z(\zeta_j^{\pm 1}). \tag{6.1.7}$$

Problem 6.1.1. Derive Equation (6.1.6) from Equation (5.1.2) using Equations (6.1.4).

If it is true that $|\gamma| \ll |\omega_r|$, then we can use Equation (5.1.16) to estimate the damping rate for $k_z > 0$:

$$\gamma \simeq -\frac{\pi}{2k_z}\sum_j \frac{\omega_j^2}{(2\pi v_j^2)^{1/2}} \exp\left[\frac{(\omega_r \pm \Omega_j)^2}{2k_z^2 v_j^2}\right]. \tag{6.1.8}$$

Thus, for Maxwellian distributions, transverse wave damping rates are proportional to sums over species that involve evaluation of the distributions at the cyclotron resonance speeds of the species.

6.2 Electromagnetic waves

Throughout this section we consider a two-species electron-proton plasma in which each species is represented by a zeroth-order Maxwellian distribution function.

6.2.1 Low frequency waves

At $\omega_r \ll \Omega_p$ and $kc/\omega_p \ll 1$, fluid theories predict that there are three normal modes of a plasma; these are usually called the magnetohydrodynamic (MHD) waves. Many fluid theories also predict that these three waves have clearly separated phase speeds at all propagation angles except at **k** parallel or perpendicular to $\mathbf{B}_o$, so that they can be labeled as fast, intermediate and slow modes (e.g. Fig. 8-2 of Boyd and Sanderson, 1969).

The Vlasov theory of long-wavelength electromagnetic waves in a homogeneous, magnetized, Maxwellian plasma yields a more detailed physical picture of MHD modes, but this additional physics renders inappropriate the fluid labels, and it becomes necessary to establish a new nomenclature. Furthermore, even though Vlasov theory yields no more than (and sometimes less than) three lightly damped waves in this regime, a single mode may have significantly different properties as a function of θ so that we believe it is necessary to identify four distinct modes in this regime. Table 6.1 compares our identification with the more traditional nomenclature based on phase speeds. The terms "quasiparallel" (q-$\|$) and "quasiperpendicular" (q-$\perp$) refer approximately to the regimes $0° \leq \theta \lesssim 30°$ and $60° \lesssim \theta \leq 90°$; the unlabeled regime between these two typically represents a transition from one type of wave to another.

As Table 6.1 indicates, the identification of low frequency waves in Vlasov theory involves three critical parameters: θ, T_e/T_p, and β. Although Vlasov theory implies two distinct kinds of sound speeds (see below), they are both proportional to the square root of plasma species temperatures. Thus since $\beta \simeq c_s^2/v_A^2$, $\beta \ll 1$ corresponds to $c_s < v_A$ and $\beta > 1$ to $v_A < c_s$.

The first low frequency wave we consider is the one that is weakly damped for almost all θ at both low and high values of β. In the long wavelength limit this wave is essentially incompressible ($C_p \ll 1$) and satisfies

$$\omega_r \simeq k_z v_A; \tag{6.2.1}$$

we call this the "Alfvén" wave. In the limit of $\mathbf{k} \times \mathbf{B}_o = 0$, this wave is left-hand circularly polarized ($P = -1$) and, if $\omega/k_z > 0$, possesses negative helicity ($\sigma = -1$). Because $v_A < c_s$ at high β, this mode has the slowest

Table 6.1. *Nomenclature for Vlasov description of low frequency waves*

Phase speeds	Weakly damped regime	Our identification
	$\beta_p << 1$	
Slow	Q-∥ if $T_e >> T_p$	Ion acoustic
Intermediate	All θ	Alfvén
Fast	Q-∥	Parallel magnetosonic
Fast	Q-⊥	Magnetosonic
	$\beta_p > 1$	
Slow	All θ	Alfvén
Intermediate	Q-∥	Parallel magnetosonic
Fast	Q-∥ if $T_e >> T_p$	Ion acoustic
Fast	Q-⊥	Magnetosonic

phase speed in this regime, but corresponds to the MHD intermediate mode at low β.

Problem 6.2.1. Numerical evaluation of the dispersion equation in a Maxwellian plasma yields $|E_y^{(1)}|^2 \gg |E_x^{(1)}|^2 \gg |E_z^{(1)}|^2$ for the Alfvén wave at $\theta \gtrsim 30°$. This implies that the dispersion equation for this wave is $D_{yy} \simeq 0$. Show that in the limits $k_y v_p/\Omega_p \ll 1$ and $|\omega|/\Omega_p \ll 1$, this equation yields Equation (6.2.1) if $v_A^2 \ll c^2$.

At $\mathbf{k} \times \mathbf{B}_o = 0$, the second wave we consider is the electrostatic root from Equation (6.1.5), i.e. the ion acoustic wave of Subsection 2.2.2. As θ increases from zero at $\beta_p \ll 1$, this mode obtains a fluctuating magnetic field such that $|\mathbf{E}^{(1)}|^2 \ll |\mathbf{B}^{(1)}|^2$. But for nonzero θ, C_p is much larger than unity and solutions of the full Vlasov dispersion equation indicate that, at low β, this fluctuating magnetic field does not significantly alter wave dispersion. Thus the phase speed is essentially the same as in the electrostatic case:

$$\frac{\omega_r}{k_z} \simeq \left(\frac{T_e + 3T_p}{m_p}\right)^{1/2} \qquad (T_e \gtrsim T_p) \qquad (6.2.2)$$

and the term "ion acoustic wave" remains an appropriate descriptor. At low β the damping is also essentially the same as in the electrostatic case; at $T_e \lesssim T_p$ the ion acoustic wave is heavily damped at all angles of propagation (Fig. 3 of Barnes, 1966), but as T_e/T_p becomes much greater than unity, the wave becomes lightly damped (Hada and Kennel, 1985). Because $c_s \ll v_A$ at $\beta \ll 1$, the ion acoustic wave corresponds to the MHD slow mode in a low β plasma; at high β it becomes the fastest wave at quasiparallel propagation.

The third wave we consider involves coupled magnetic and density fluctuations such that $C_p \sim 1$ (e.g. Fig. 11 of Gary, 1986); we term this the "magnetosonic wave." At $\beta < 1$ and quasiperpendicular propagation, this wave satisfies

$$\omega_r^2 \simeq k^2(v_A^2 + c_{s\perp}^2 \sin^2\theta) \tag{6.2.3}$$

where

$$c_{s\perp}^2 = 2\,\frac{T_p + T_e}{m_p},$$

the square of the ion sound speed at propagation perpendicular to $\mathbf{B}_o$. At quasiperpendicular propagation, this wave is weakly damped for both low and high β (Hada and Kennel, 1985), and corresponds to the fast MHD mode at all β.

Problem 6.2.2. Numerical evaluation of the dispersion equation in a Maxwellian plasma yields $|E_x^{(1)}|^2 \gg |E_y^{(1)}|^2 \gg |E_z^{(1)}|^2$ for the magnetosonic wave at $\theta \gtrsim 30°$. This implies that the dispersion equation for this wave is $D_{xx} \simeq 0$. Show that in the limits $k_y v_p/\Omega_p \ll 1$ and $|\omega|/\Omega_p \ll 1$, this equation yields Equation (6.2.3) if $v_A^2 \ll c^2$.

The fourth low frequency wave we identify is noncompressive, satisfies $\omega_r \simeq kv_A$, and is weakly damped at quasiparallel propagation. Unlike the Alfvén wave, however, in the limit of $\mathbf{k} \times \mathbf{B}_o = 0$, this wave is right-hand circularly polarized ($P = +1$) and, if $\omega/k_z > 0$, possesses positive helicity ($\sigma = +1$). At low β this mode has the fastest phase speed and represents the extension of the magnetosonic wave to quasiparallel propagation; at high β this wave has an intermediate phase speed. In order to retain the traditional low-β association of this mode with the magnetosonic wave, we term this the "parallel magnetosonic wave."

Figure 6.1 illustrates the phase speeds ω_r/kv_A of low frequency waves as a function of wave propagation direction for six pairs of the parameter set $\{\beta_p, T_e/T_p\}$. At low β_p, there are three waves with clearly separated phase speeds, the increase in damping of the ion acoustic wave with decreasing T_e/T_p is evident, and the ion acoustic wave is the mode of slowest phase speed. At high β_p a greater fraction of wavenumber space corresponds to heavily damped modes, the phase speeds are less clearly separated at oblique propagation, and the mode of slowest phase speed is the Alfvén wave. From Equation (6.2.2) it follows that the condition that separates the low and high β propagation regimes is $\beta \simeq 2/(3 + T_e/T_p)$.

The properties of five transport ratios for each of the low frequency waves at $T_e = T_p$ are given in Table 6.2, and are illustrated in Figure 6.2. The

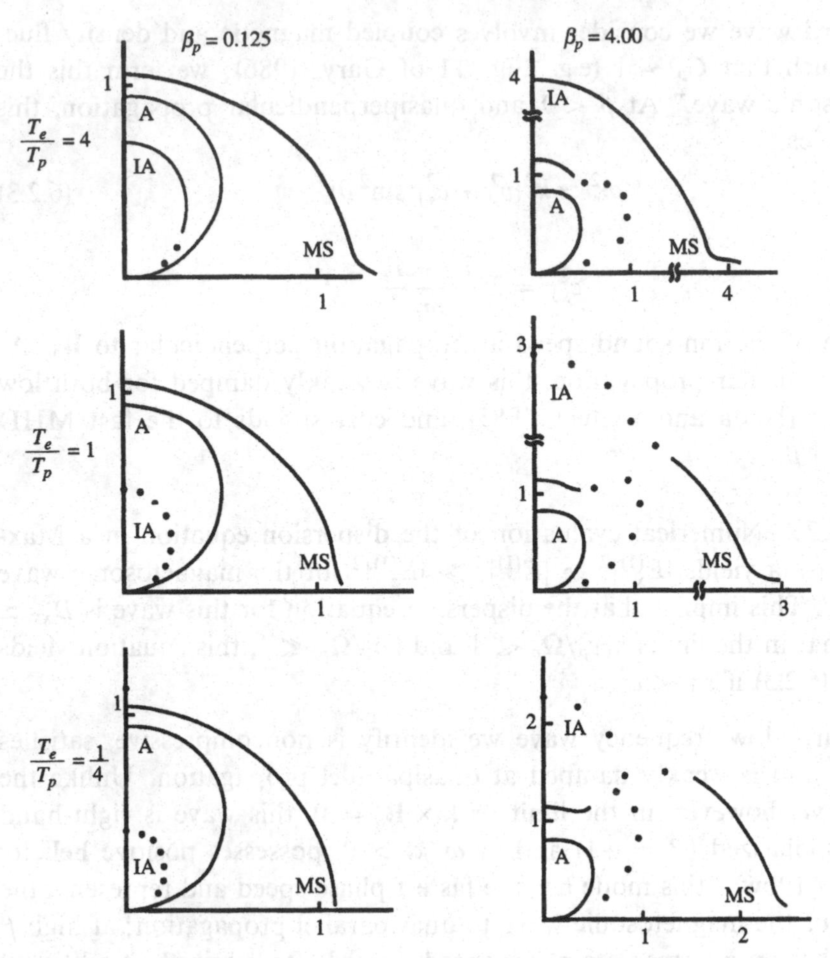

Fig. 6.1 The phase speed ω_r/kv_A of low frequency waves in a plasma of Maxwellian electrons and protons for six values of the parameter pair $\beta_p, T_e/T_p$. The direction of wave propagation θ is measured with respect to the vertical axis, which represents the direction of $\mathbf{B}_o$. Solid lines denote weakly damped modes, $|\gamma| < |\omega_r|/2\pi$, whereas dotted lines indicate heavily damped fluctuations, $\gamma < -|\omega_r|/2\pi$. The label "IA" indicates an ion acoustic wave ($C_p >> 1$) at small θ; "A" denotes an Alfvén wave ($C_p << 1$) at all angles of propagation, and "MS" labels a magnetosonic wave ($C_p \sim 1$) at $\theta \gtrsim 60°$ (From Fig. 1 of Gary, 1992). Here, and for all the figures in this chapter unless otherwise noted, we consider an electron-proton plasma with $T_e = T_p$.

same properties at $T_e = T_p/4$ and $T_e = 4T_p$ are summarized in Tables I and II of Gary and Winske (1992). This figure indicates that the the parallel compressibility and the Alfvén ratio, as well as the compressibility, are the quantities most likely to permit differentiation between the obliquely propagating magnetosonic and Alfvén waves.

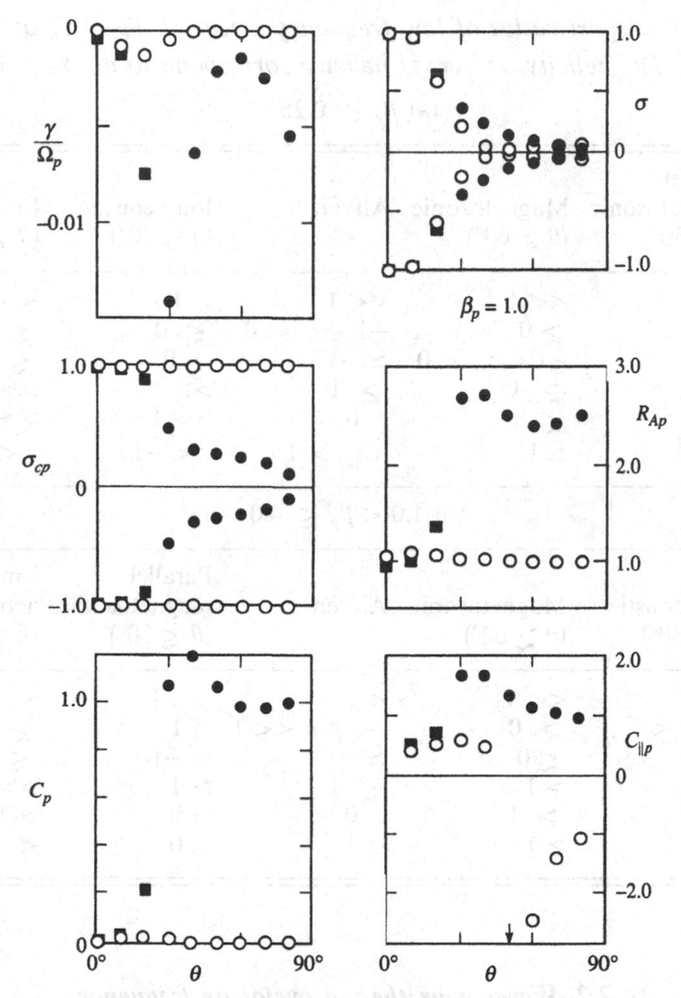

Fig. 6.2 The damping rate and five transport ratios of the lightly damped regime of the parallel magnetosonic wave (solid squares), the (oblique) magnetosonic wave (solid dots), and the Alfvén wave (open dots) as functions of propagation angle for $\beta_p = 1.0$. Here $kc/\omega_p = 0.10$. At $\omega_r/k_z > 0$, the lightly damped magnetosonic waves both have $\sigma > 0$ and $\sigma_{cp} < 0$, and the Alfvén wave has $\sigma < 0$ and $\sigma_{cp} < 0$ at small θ.

Problem 6.2.3. Consider the Alfvén wave propagating at $\theta \gtrsim 30°$ and $\omega_r \ll \Omega_p$. Assuming, as in Problem 6.2.1, that $E_x^{(1)} = E_z^{(1)} = 0$, show that the proton compressibility can be written as

$$C_p \simeq \left(\frac{k_z c}{\omega_p}\right)^2 \left(\frac{k_y c}{\omega_p}\right)^2$$

Table 6.2. *Transport ratios of low frequency waves at $T_e = T_p$ at $kc/\omega_p = 0.10$. The helicity and cross helicity correspond to $\omega_r/k_z > 0$.*

(a) $\beta_p \lesssim 0.25$

	Parallel Magnetosonic ($\theta \lesssim 30^o$)	Magnetosonic ($\theta \gtrsim 60^o$)	Alfvén	Ion acoustic ($\theta \lesssim 30^o$)	Ion acoustic ($\theta \gtrsim 60^o$)
$\lvert\gamma\rvert/\omega_r$	$<<1$	$<<1$	$<<1$	~ 1	~ 1
σ	≤ 1	$\gtrsim 0$	$-1 \leq \sigma \lesssim 0$	$\lesssim 0$	$\lesssim 0$
σ_{cp}	$\simeq -1$	$-1 < \sigma_{cp} < 0$	$\simeq -1$	$\lesssim 0$	$\lesssim 0$
R_{Ap}	$\lesssim 1$	$\gtrsim 1$	$\gtrsim 1$	$>> 1$	$>> 1$
C_p	$\simeq 0$	$\lesssim 1$	$\simeq 0$	$>> 1$	$>> 1$
$C_{\parallel p}$	$\simeq 1$	$\simeq 1$	$\lvert C_{\parallel p}\rvert > 1$	$<< -1$	$<< -1$

(b) $1.0 < \beta_p \lesssim 4.0$

	Ion acoustic ($\theta \lesssim 30^o$)	Magnetosonic ($\theta \gtrsim 60^o$)	Alfvén	Parallel magnetosonic ($\theta \lesssim 30^o$)	Ion acoustic ($\theta \gtrsim 60^o$)
$\lvert\gamma\rvert/\omega_r$	~ 1	$<<1$	$<<1$	$<<1$	~ 1
σ	$0 < \sigma < 1$	$\gtrsim 0$	$-1 \leq \sigma << 1$	≤ 1	$\gtrsim 0$
σ_{cp}	$\lesssim 0$	$\lesssim 0$	$\simeq -1$	$\simeq -1$	$\lesssim 0$
R_{Ap}	$>> 1$	> 1	$\lesssim 1$	> 1	$>> 1$
C_p	$>> 1$	$\gtrsim 1$	$\simeq 0$	$\simeq 0$	> 1
$C_{\parallel p}$	$>> 1$	≥ 1	< 1	$\simeq 0$	< -1

6.2.2 *Waves near the ion cyclotron frequency*

As wavenumber and frequency increase, the low frequency waves of Vlasov theory exhibit different behaviors; these are illustrated in Figure 6.3, Figure 6.4, and Figure 6.5.

The obliquely propagating magnetosonic wave is characterized by $C_p \sim 1$ at both low and high β_p and an increasing phase speed as k increases and ω_r passes through Ω_p. The absolute value of the damping ratio gradually increases as the wavenumber and frequency increase, but in this frequency regime the damping becomes appreciable near $\omega_r \sim \Omega_p$ only at relatively high β_p, as illustrated in Figure 6.3.

Although the parallel magnetosonic wave is noncompressive at $\mathbf{k} \times \mathbf{B}_o = 0$, it also exhibits a phase speed that increases with wavenumber through the proton cyclotron resonance. Like the compressive magnetosonic wave,

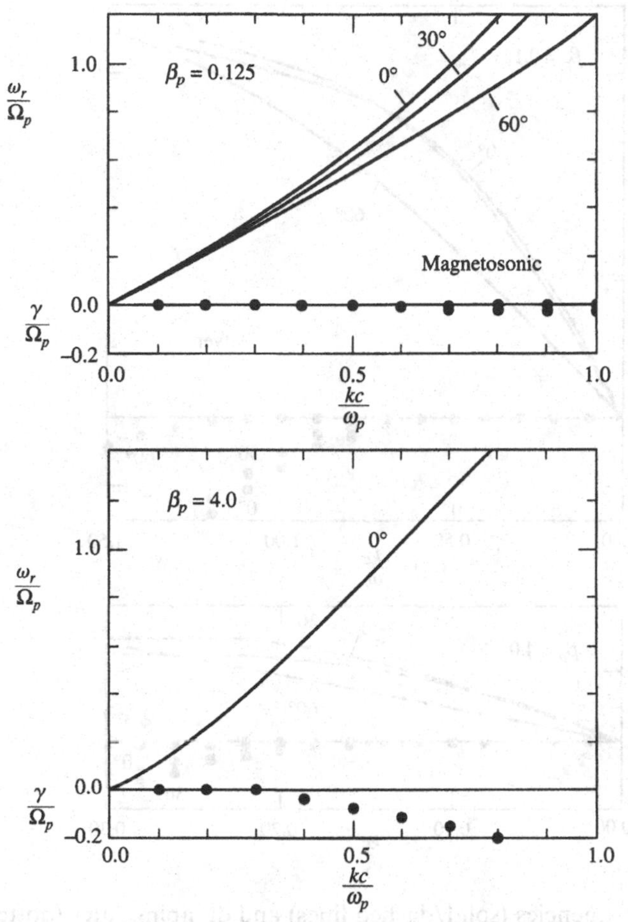

Fig. 6.3 The real frequencies (solid lines) and damping rates (dotted lines) as functions of wavenumber for the mode that evolves from the parallel magnetosonic wave at three values of θ for $\beta_p = 0.125$ and $\theta = 0°$ at $\beta_p = 4.0$. The corresponding fluctuations at $\beta_p = 4.0$ and $\theta \gtrsim 30°$ are heavily damped here. Here $v_A/c = 10^{-4}$.

the parallel magnetosonic wave is lightly damped at low β_p and exhibits increased damping with wavenumber at high β_p.

The left-hand polarization of the Alfvén wave at $\mathbf{k} \times \mathbf{B}_o = 0$ implies that this mode resonates with ions as k and ω_r increase; as Figure 6.4 shows, this resonance bounds ω_r well below the proton cyclotron frequency at all propagation angles and leads to cyclotron damping at short wavelengths at almost all θ values. To describe this wave over its full frequency range, we use the term "Alfvén/ion cyclotron" wave to avoid confusion with the electrostatic ion Bernstein waves at $\omega_r > \Omega_p$ (see Subsection 2.3.1).

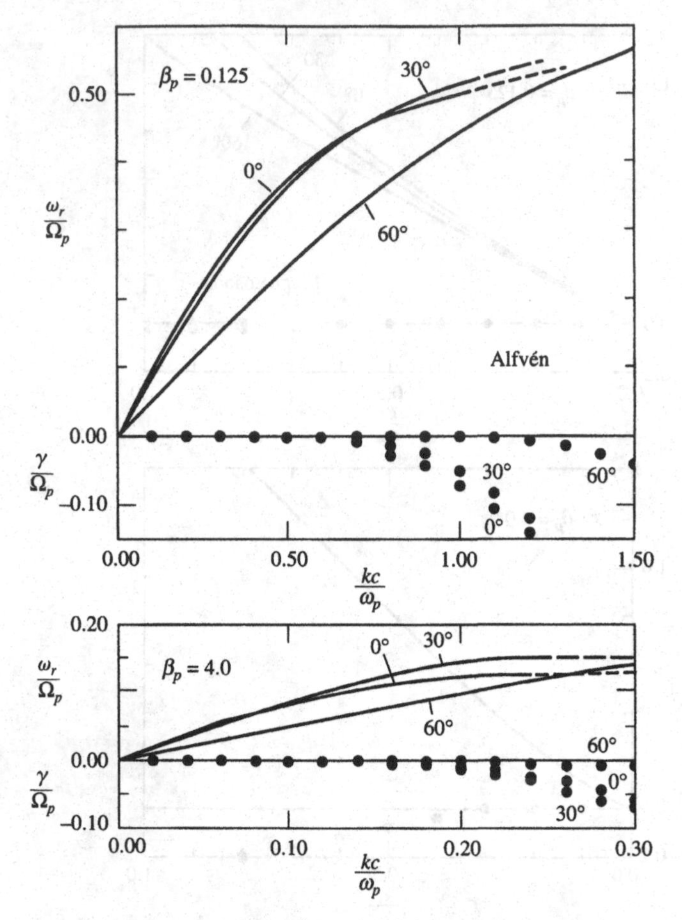

Fig. 6.4 The real frequencies (solid/dashed lines) and damping rates (dotted lines) as functions of wavenumber for the Alfvén wave at three values of θ for $\beta_p = 0.125$ and $\beta_p = 4.0$. A dashed line indicates that the mode is heavily damped ($\gamma < -|\omega_r|/2\pi$). Here $v_A/c = 10^{-4}$.

In contrast, the ion acoustic wave is electrostatic at $\mathbf{k} \times \mathbf{B}_o = 0$ and at $T_e \gg T_p$ propagates through the proton cyclotron resonance, as shown in Figure 6.5. At oblique propagation, however, the fluctuating fields of this wave become primarily transverse and the real part of the polarization becomes of the order of -1 as the wavenumber increases. This implies a proton cyclotron resonance similar to that of the Alfvén wave and, at low β_p and oblique propagation, the ion acoustic wave experiences a similar damping and dispersion. At high β_p and oblique propagation, the real part of the polarization of the ion acoustic wave becomes of order $+1$, so that this mode evolves into the magnetosonic wave.

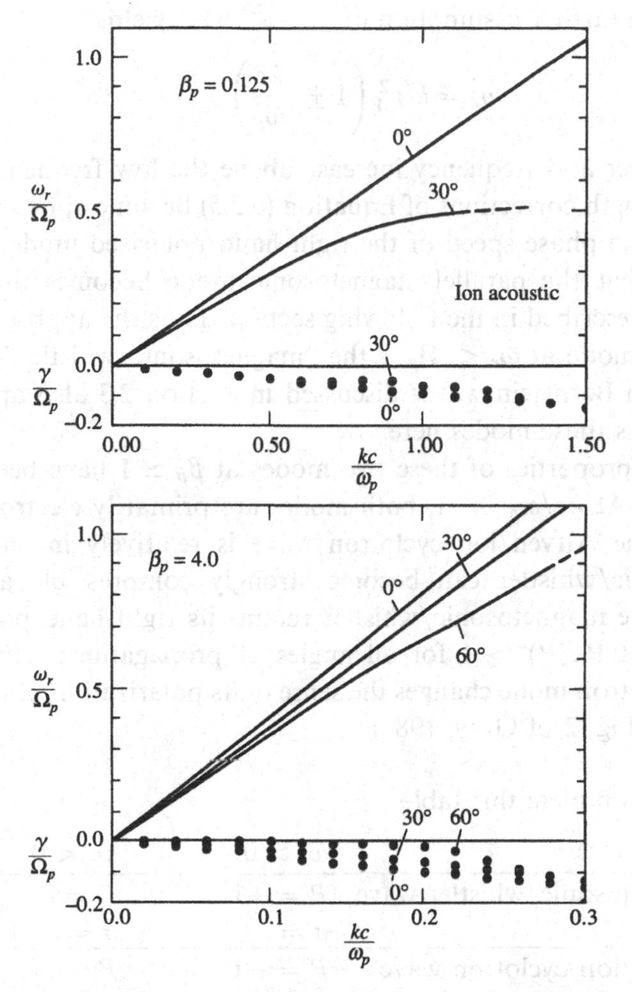

Fig. 6.5 The real frequencies (solid/dashed lines) and damping rates (dotted lines) as functions of wavenumber for the mode that evolves from the ion acoustic wave at several values of θ for $\beta_p = 0.125$ and $\beta_p = 4.0$. The corresponding wave at $\beta_p = 0.125$ and $\theta = 60°$ is heavily damped here. A dashed line indicates that the mode is heavily damped ($\gamma < -|\omega_r|/2\pi$). Here $v_A/c = 10^{-4}$ and $T_e = 4T_p$.

At $ka_p \ll 1$ and $\mathbf{k} \times \mathbf{B}_o = 0$, both the Alfvén wave and the parallel magnetosonic wave are nonresonant; i.e. $|\zeta_j^{\pm 1}| \gg 1$ for both species. In this limit, Equation (6.1.6) yields the cold plasma dispersion equation

$$\omega^2 - k^2c^2 - \sum_j \frac{\omega\omega_j^2}{\omega \pm \Omega_j} = 0, \tag{6.2.4}$$

which, under the further assumption of $|\omega| \ll |\Omega_j|$, yields

$$\omega_r^2 = k^2 v_A^2 \left(1 \pm \frac{kc}{\omega_p} \right). \tag{6.2.5}$$

As wavenumber and frequency increase above the low frequency regime, the finite wavelength corrections of Equation (6.2.5) become apparent. Figure 6.3 shows that the phase speed of the right-hand polarized mode continues to increase so that the parallel magnetosonic mode becomes the whistler mode, which is described in the following section. Thus the appropriate term to describe this mode at $\omega_r \leq \Omega_p$ is the "magnetosonic/whistler" wave. At $\omega_r > \Omega_p$, the ion Bernstein waves discussed in Section 2.3 also appear, but we do not discuss these modes here.

Other Vlasov properties of these two modes at $\beta_p \simeq 1$ have been studied by Gary (1986). At $kc/\omega_p \ll 1$, both modes are primarily electromagnetic, and although the Alfvén/ion cyclotron wave is relatively incompressible, the magnetosonic/whistler can become strongly compressible at oblique propagation. The magnetosonic/whistler retains its right-hand polarization in the sense that $\mathrm{Re}(P) \gtrsim 1$ for all angles of propagation; however, the Alfvén/ion cyclotron mode changes the sense of its polarization at sufficiently large β_p and θ (Fig. 2 of Gary, 1986).

Problem 6.2.4. Complete this table.

$\mathbf{k} \times \mathbf{B}_o = 0$	$\omega_r > 0$	$\omega_r < 0$
Magnetosonic/whistler wave	$P = +1$	$P =$
	$\sigma =$	$\sigma =$
Alfvén/ion cyclotron wave	$P = -1$	$P =$
	$\sigma = -1$	$\sigma =$

Near and above the proton cyclotron frequency, the four transport ratios of Subsection 5.2.3 that involve plasma quantities become less useful as mode identifiers. Present technology requires a few seconds to complete an accurate spacecraft measurement of electron or ion distribution functions; because this time is of the order of a proton gyroperiod in many space plasmas, it is difficult to obtain plasma densities and flow velocities rapidly enough to construct the associated transport ratios at $\omega \gtrsim \Omega_p$. Measurements of magnetic and electric fields are, of course, possible at much faster time scales, so transport ratios involving only $\delta\mathbf{B}$ and $\delta\mathbf{E}$ are of interest at higher frequencies.

Figure 6.6 illustrates the electric/magnetic field energy ratio as a function of k for two modes that can propagate through the proton cyclotron

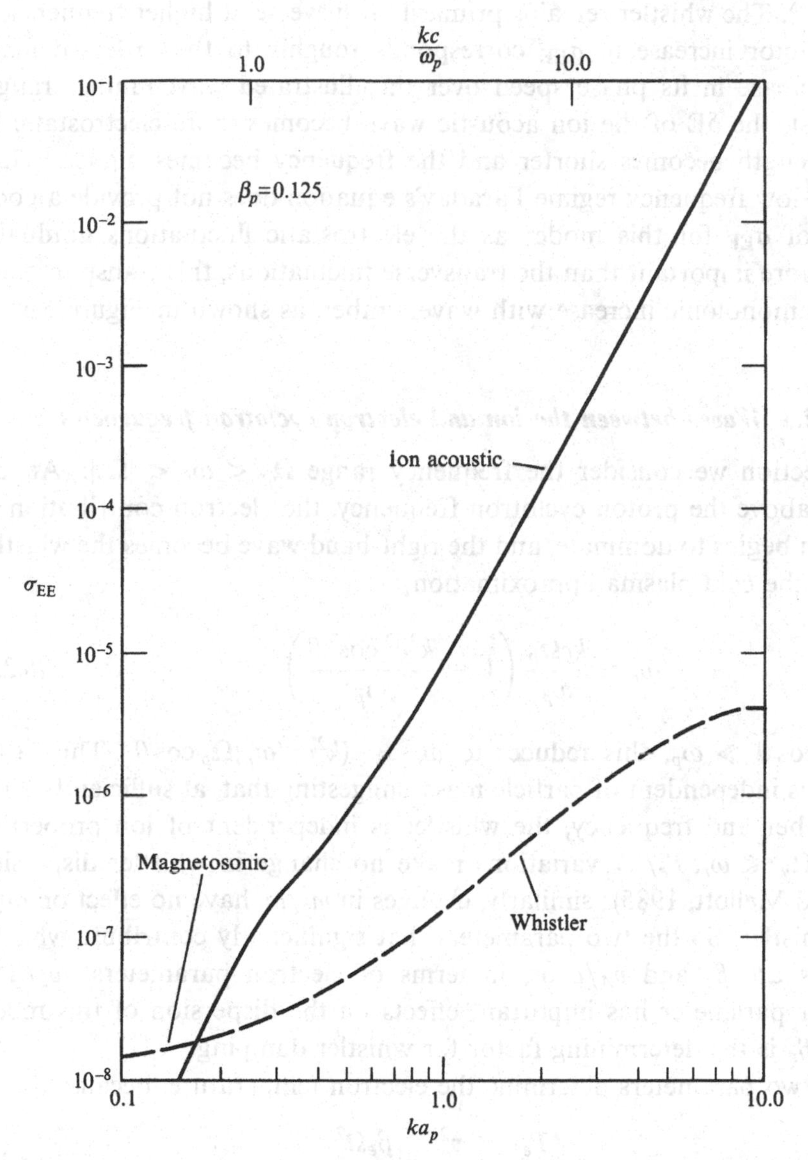

Fig. 6.6 The electric/magnetic field energy ratio as a function of wavenumber for two electromagnetic modes at $\beta_p = 0.125$ and $v_A/c = 10^{-4}$. The magnetosonic/whistler wave propagates at $\theta = 30°$ in a $T_e = T_p$ plasma; the ion acoustic wave propagates at $\theta = 10°$ in a $T_e = 4T_p$ plasma.

resonance to higher frequencies: the magnetosonic/whistler wave and the ion acoustic wave at quasiparallel propagation. At long wavelengths and the parameters chosen here, the fluctuating electric fields of both modes are primarily transverse to **k**, so that, from Faraday's equation and $v_A/c = 10^{-4}$,

$\sigma_{\mathbf{EE}} \sim 10^{-8}$. The whistler remains primarily transverse at higher frequencies; the 10^2 factor increase in $\sigma_{\mathbf{EE}}$ corresponds roughly to the order of magnitude increase in its phase speed over the illustrated wavenumber range. In contrast, the $\delta\mathbf{E}$ of the ion acoustic wave becomes more electrostatic as the wavelength becomes shorter and the frequency becomes higher. Thus above the low frequency regime Faraday's equation does not provide a good estimate of $\sigma_{\mathbf{EE}}$ for this mode; as the electrostatic fluctuations gradually become more important than the transverse fluctuations, this transport ratio exhibits a monotonic increase with wavenumber, as shown in Figure 6.6.

6.2.3 Waves between the ion and electron cyclotron frequencies

In this section we consider the frequency range $\Omega_p < \omega_r < |\Omega_e|$. At frequencies above the proton cyclotron frequency, the electron contribution to dispersion begins to dominate, and the right-hand wave becomes the whistler mode. In the cold plasma approximation,

$$\omega_r \simeq \frac{kc\Omega_p}{\omega_p}\left(1 + \frac{k^2c^2\cos^2\theta}{\omega_p^2}\right)^{1/2}. \tag{6.2.6}$$

At $kc\cos\theta \gg \omega_p$, this reduces to $\omega_r \simeq (k^2c^2/\omega_p^2)\Omega_p\cos\theta$. This latter equation is independent of particle mass, suggesting that, at sufficiently large wavenumber and frequency, the whistler is independent of ion properties. Thus at $\Omega_p \ll \omega_r$, T_e/T_p variations make no change in whistler dispersion (Gary and Mellott, 1985); similarly, changes in m_e/m_i have no effect on $\omega(\mathbf{k})$ of the whistler. So the two parameters that significantly contribute whistler properties are β_e and v_A/c or, in terms of electron parameters, $\omega_e/|\Omega_e|$. The latter parameter has important effects on the dispersion of this mode, whereas β_e is the determining factor for whistler damping.

These two parameters determine the electron temperature, because

$$\frac{T_e}{m_ec^2} = \frac{v_e^2}{c^2} = \frac{\beta_e\Omega_e^2}{2\omega_e^2}. \tag{6.2.7}$$

Note that, if $\Omega_e^2 \gtrsim \omega_e^2$ and the nonrelativistic theory used here is to remain valid, this equation requires $\beta_e \ll 1$.

As ω_r increases further, the whistler undergoes a resonance and Equation (6.2.6) is no longer valid. In this frequency regime, the dispersion properties depend on $\omega_e/|\Omega_e|$, and the cases of large and small values of this parameter must be distinguished.

We first consider the case $\omega_e/|\Omega_e| > 1$. Figure 6.7 illustrates representative

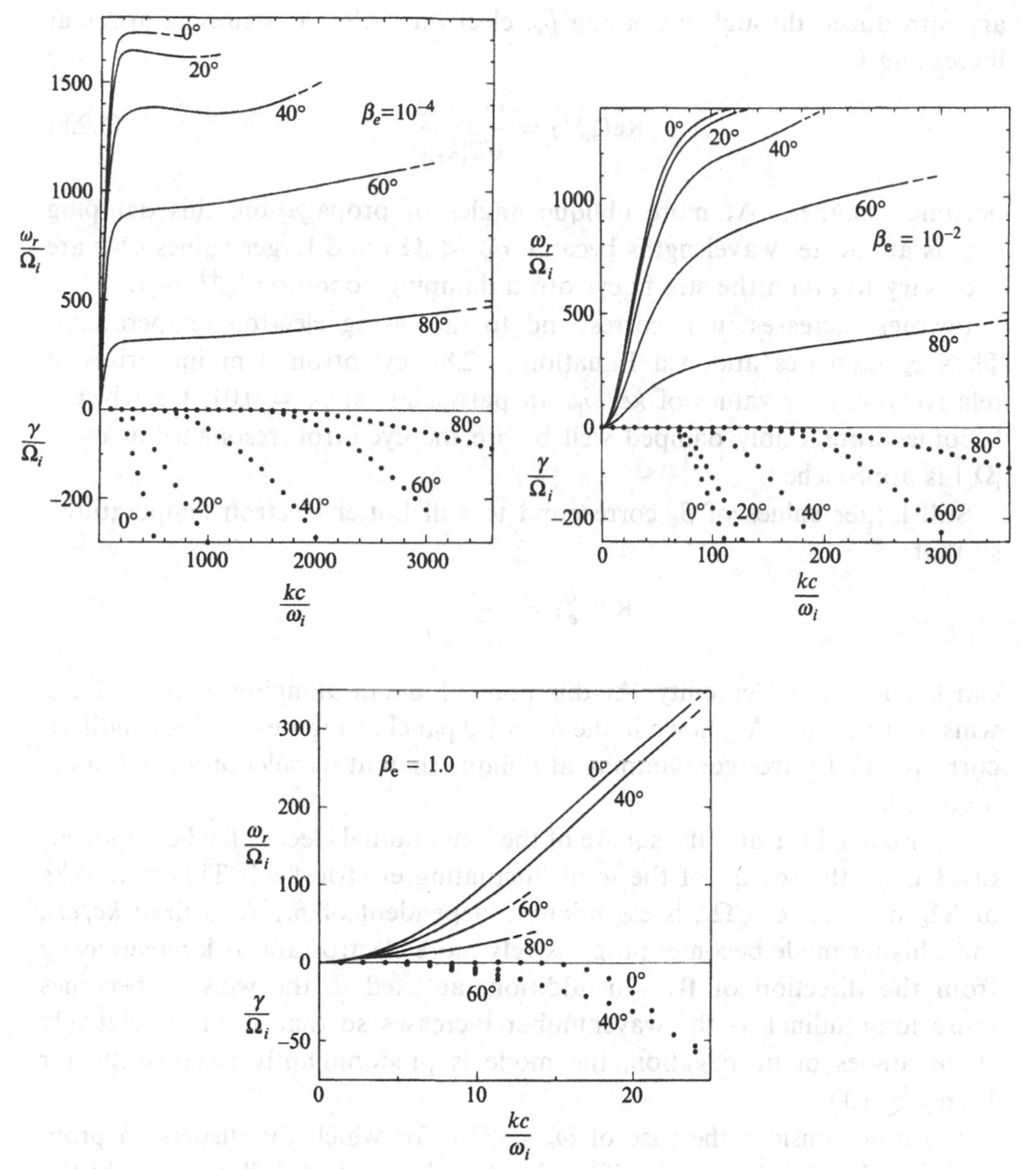

Fig. 6.7 The real frequencies (solid/dashed lines) and damping rates (dotted lines) as functions of wavenumber for the whistler/electron cyclotron wave at several different values of θ at $\omega_e/|\Omega_e| = 10.0$ and three different values of β_e. A dashed line indicates that the mode is heavily damped ($\gamma < -|\omega_r|/2\pi$).

dispersion properties at $\omega_e \gg |\Omega_e|$ for three different values of β_e. In the $\beta_e = 0$ cold plasma limit, the increase in ω_r with k described by Equation (6.2.6) is terminated by the resonance at $|\Omega_e| \cos\theta$. At these frequencies, we call the right-hand polarized wave the "electron cyclotron wave." As thermal effects are introduced through increasing β_e, electron cyclotron damping arises at increasing k as

$$\mathrm{Re}(\zeta_e^{+1}) = \frac{\omega_r + \Omega_e}{\sqrt{2}|k_z|v_e} \tag{6.2.8}$$

becomes smaller. At more oblique angles of propagation, this damping begins at shorter wavelengths because $\omega_r \ll |\Omega_e|$ and larger values of k are necessary to attain the strong cyclotron damping condition $|\zeta_e^{+1}| \sim 1$.

Further increases in β correspond to increasing electron temperatures. Thus v_e increases and, via Equation (6.2.8), cyclotron damping arises at relatively smaller values of kc/ω_p. In particular, at $\beta_e = 0.01$, the whistler becomes appreciably damped well before the cyclotron resonance at $\omega_r = |\Omega_e|$ is approached.

Still larger values of β_e correspond to still hotter electron temperatures, so that

$$\mathrm{Re}(\zeta_e^{0}) = \frac{\omega_r}{\sqrt{2}|k_z|v_e}$$

can become of order unity. At this point, Landau damping of the oblique whistler can arise. As shown in the $\beta_e = 1.0$ panel of Figure 6.7, this condition corresponds to stronger damping at oblique than at parallel propagation for a given k.

Figure 6.8 illustrates the square of the longitudinal electric field component divided by the square of the total fluctuating electric field. This ratio, like ω_r/Ω_p but unlike γ/Ω_p, is essentially independent of β_e. At a fixed kc/ω_p, the whistler mode becomes progressively more electrostatic as $\mathbf{k}$ moves away from the direction of $\mathbf{B}_o$. In addition, at fixed θ, the whistler becomes more longitudinal as the wavenumber increases so that, even at relatively small angles of propagation, the mode is predominantly electrostatic for $kc/\omega_p \gtrsim 100$.

We now consider the case of $\omega_e < |\Omega_e|$, for which the dispersion properties of the whistler are significantly altered. Figure 6.9 illustrates whistler dispersion at $\omega_e \ll |\Omega_e|$ for two different values of β_e. From Equation (6.2.7), the $\beta_e = 1.0$ panel of Figure 6.7 and the $\beta_e = 10^{-4}$ panel of Figure 6.9 correspond to the same electron temperature; comparison of these two panels clearly indicates that it is β_e, not the electron temperature, which determines whistler damping. Although ω_r at $\mathbf{k} \times \mathbf{B}_o = 0$ shows dispersion properties similar to those at $\omega_e > |\Omega_e|$, this corresponds to a set of measure zero, because the effects of the plasma frequency cutoff appear at arbitrarily small nonzero θ.

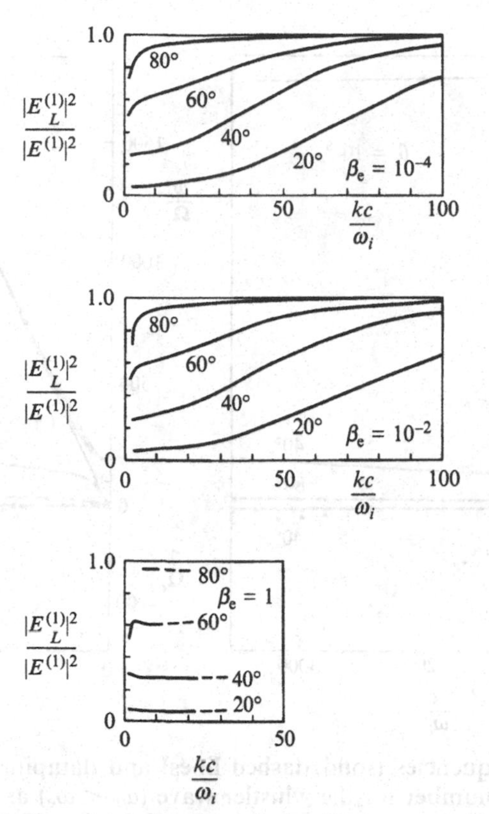

Fig. 6.8 The ratio of the square of the longitudinal component of the fluctuating electric field to the total fluctuating electric field as functions of wavenumber for the whistler/electron cyclotron wave at several different values of θ for three different values of β_e at $\omega_e/|\Omega_e| = 10.0$. A dashed line indicates that the mode is heavily damped ($\gamma < -|\omega_r|/2\pi$) (From Fig. 4 of Tokar and Gary, 1985).

Thus, in this case, the lower frequency branch of the whistler becomes damped as the frequency approaches the electron plasma resonance $\omega_r \sim \omega_e \cos\theta$, while the upper frequency branch, called the Z mode, propagates relatively undamped between ω_e and $|\Omega_e|$. Once again, increasing β_e increases both cyclotron and Landau damping and reduces the wavenumber range of both whistler and Z mode propagation.

Figure 6.10 shows that in this parameter range as well $|E_L^{(1)}|^2/|E^{(1)}|^2$ is independent of β_e. The plots of this quantity vs wavenumber for the whistler are similar to those at $\omega_e > |\Omega_e|$, except that the whistler becomes predominantly longitudinal at considerably smaller values of kc/ω_p. Thus at $kc/\omega_e \gtrsim 1$ and oblique propagation, the mode is known as the "electrostatic whistler." In contrast, the Z mode is predominantly longitudinal at relatively

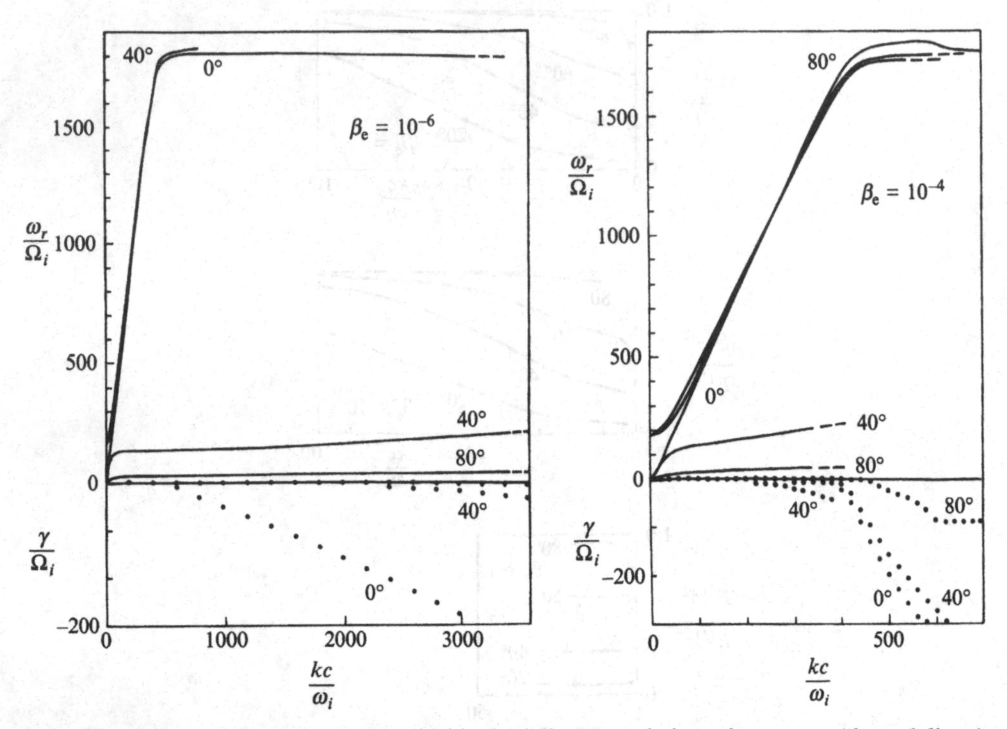

Fig. 6.9 The real frequencies (solid/dashed lines) and damping rates (dotted lines) as functions of wavenumber for the whistler wave ($\omega_r < \omega_e$) and Z mode ($\omega_r > \omega_e$) at $\theta = 0°$, 40° and 80° for two different values of β_e at $\omega_e/|\Omega_e| = 0.10$. A dashed line indicates that the mode is heavily damped ($\gamma < -|\omega_r|/2\pi$). At $\beta_e = 10^{-6}$, the Z mode plot at 40° is incomplete and the Z mode plot at 80° is missing because of numerical problems in tracking this mode.

small θ and relatively small kc/ω_p, reflecting the electrostatic nature of the $\omega_r = \omega_e$ cutoff. As the wavenumber increases, this branch becomes less longitudinal so that for $kc/\omega_p \gtrsim 50$ the mode is almostly fully transverse. Although it is not shown in the figure, the Z mode returns to predominantly longitudinal character at wavelengths sufficiently short that ω_r approaches the electron cyclotron resonance.

6.2.4 Waves above the electron cyclotron frequency

At $\mathbf{k} \times \mathbf{B}_o = 0$, the left-hand electromagnetic wave undergoes an ion cyclotron resonance at $\omega_r < \Omega_p$, and the right-hand wave undergoes electron cyclotron damping at $\omega_r < |\Omega_e|$. However, for ω_r at and above the electron cyclotron frequency, new modes with both polarizations appear at small values of k.

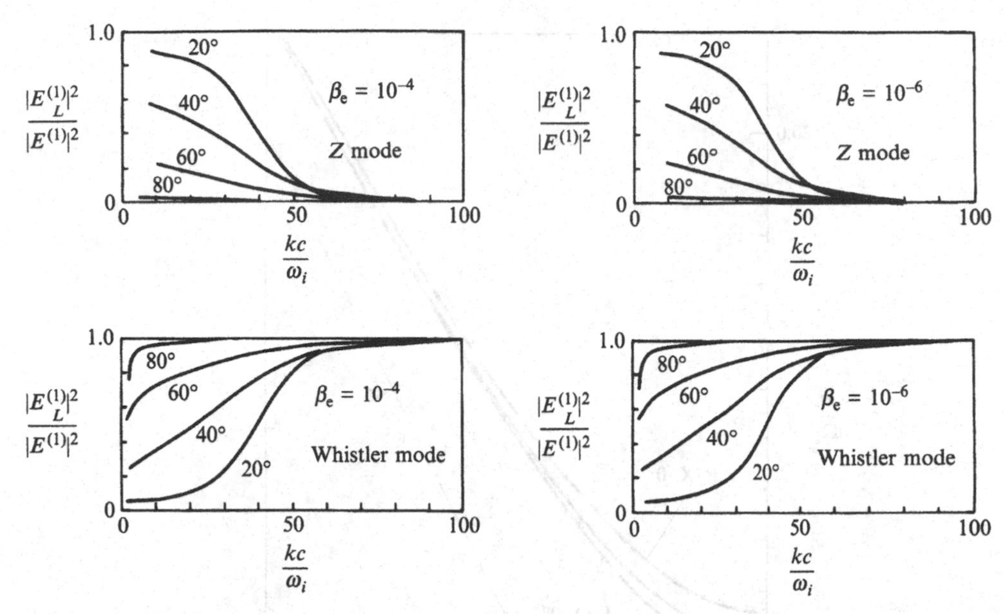

Fig. 6.10 The ratio of the square of the longitudinal component of the fluctuating electric field to the total fluctuating electric field as a function of wavenumber for the whistler wave and the Z mode at several different values of θ for two different values of β_e at $\omega_e/|\Omega_e| = 0.10$ (The bottom two panels are from Fig. 5 of Tokar and Gary, 1985).

The long wavelength cutoffs of these waves are obtained directly from the cold plasma dispersion equation (6.2.4); they are

$$\omega_{cut} = \frac{1}{2}[\mp\Omega_e + (\Omega_e^2 + 4\omega_e^2)^{1/2}]. \tag{6.2.9}$$

At short wavelengths, it is also clear from Equation (6.2.4) that both modes become light waves, $\omega_r \simeq kc$.

At lower frequencies, the dispersion properties are functions of $\omega_e/|\Omega_e|$, and we again must consider two distinct cases.

Typical dispersion properties at $\omega_e/|\Omega_e| \gg 1$ are shown in Figure 6.11. As Equation (6.2.9) implies, the cutoff frequency of the left-hand mode lies somewhat below that of the right-hand mode. As θ increases away from $0°$, the left-hand mode becomes elliptically polarized ($-1 < \mathrm{Re}(P) < 0$), the cutoff frequency increases toward ω_e and, at $kc/\omega_e \lesssim 1, \mathbf{E}^{(1)}$ points primarily in the direction of $\mathbf{B}_o$.

In the limit of perpendicular propagation, $\mathbf{E}^{(1)}$ becomes parallel to the magnetic field and the dispersion equation becomes that of an electromag-

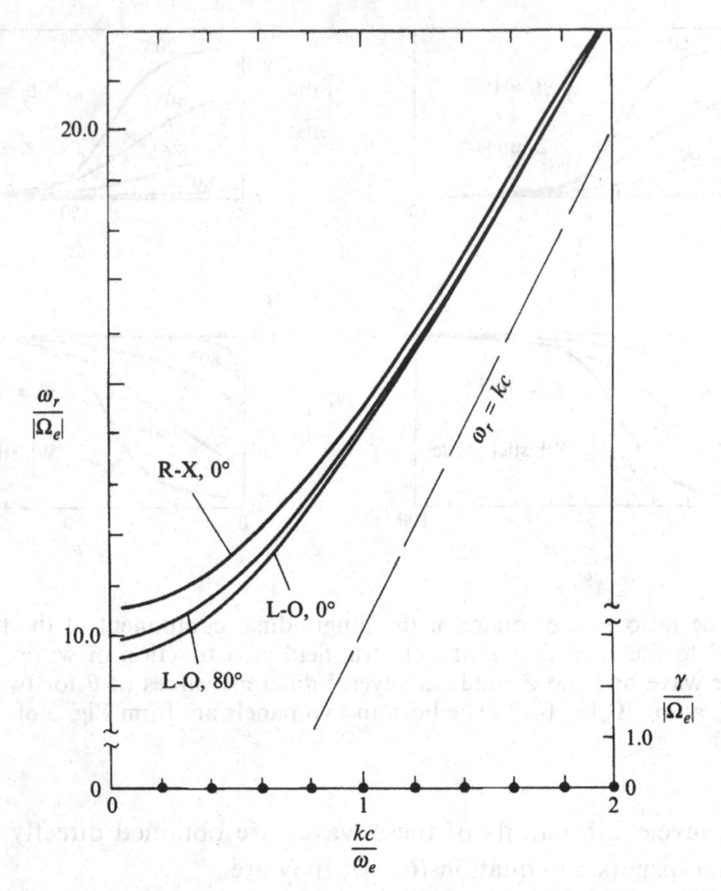

Fig. 6.11 The real frequencies (solid lines) and damping rates (dotted lines) of the R-X and L-O waves as functions of wavenumber. The L-O mode is shown at $\theta = 0°$ and 80°; the R-X mode is shown at $\theta = 0°$, but its dispersion properties at $\theta = 80°$ are virtually indistinguishable from the case of parallel propagation. Here $\beta_e = 1.0$ and $\omega_e = 10|\Omega_e|$; the ions are ignored.

netic wave in an unmagnetized plasma:

$$\omega_r^2 = \omega_e^2 + k^2c^2.$$

At $\theta = 90°$, this wave is called the "ordinary mode;" we use the term "L-O mode" to represent the wave at arbitrary directions of propagation.

The cutoff frequency of the right-hand mode lies somewhat above ω_e at $\omega_e/|\Omega_e| \gg 1$, and does not change significantly as θ increases. At oblique propagation the mode becomes right-hand elliptically polarized

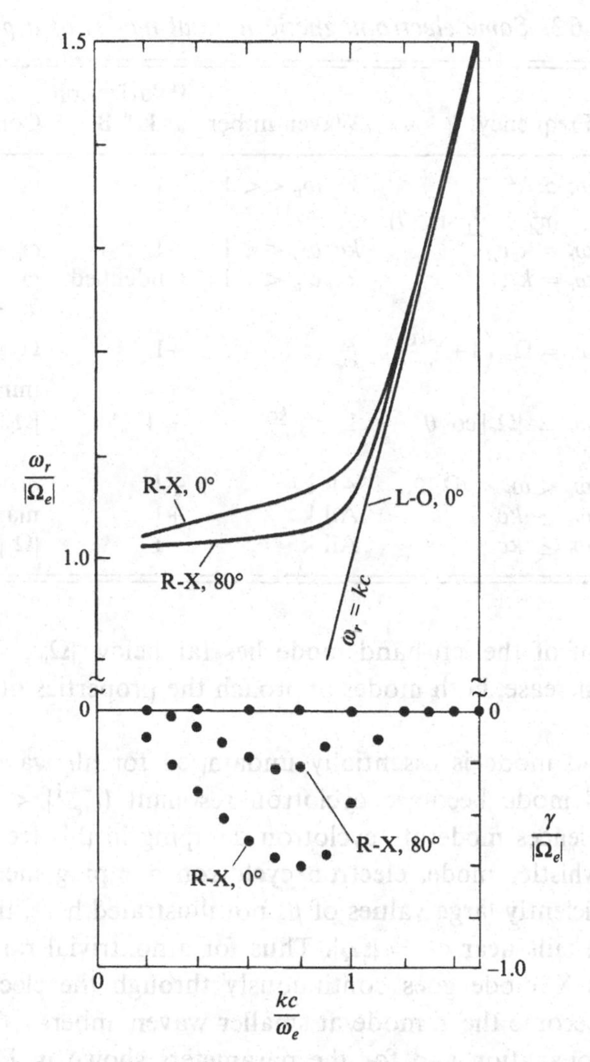

Fig. 6.12 The real frequencies (solid lines) and damping rates (dotted lines) of the R-X and L-O waves as functions of wavenumber. The R-X mode is shown at $\theta = 0°$ and 80°; the L-O mode is shown at $\theta = 0°$, but its dispersion properties at $\theta = 80°$ are virtually indistinguishable from the case of parallel propagation. Here $\beta_e = 10^{-4}$ and $\omega_e = 0.10|\Omega_e|$; the ions are ignored.

($0 < \text{Re}(P) < 1$) and has $\mathbf{E}^{(1)}$ predominantly perpendicular to $\mathbf{B}_o$. In the limit of perpendicular propagation this mode is called the "extraordinary mode;" we denote it as the "R-X mode."

Typical dispersion properties at $\omega_e/|\Omega_e| \ll 1$ are shown in Figure 6.12. As Equation (6.2.9) implies, at $\mathbf{k} \times \mathbf{B}_o = 0$ the cutoff frequency of the right-hand polarized mode is somewhat above the electron cyclotron frequency,

Table 6.3. *Some electromagnetic normal modes of a plasma*

Name	Frequency	Wavenumber	Polarization at **k** ∥ **B**	Conditions
Magnetosonic	$\omega_r^2 \simeq k^2 (v_A^2 + c_{s\perp}^2 \sin^2\theta)$	$kc/\omega_p << 1$	+1	$\omega_r < \Omega_p$
Alfvén	$\omega_r = k_z v_A$	$kc/\omega_p << 1$	−1	$\omega_r << \Omega_p$
Ion acoustic	$\omega_r = k_z c_s$	$kc/\omega_p << 1$	Undefined	$\omega_r << \Omega_p$, $T_p << T_e$
Whistler	$\omega_r = \Omega_p \left(1+\frac{c^2kk_z}{\omega_p^2}\right)$	$\frac{kc}{\omega_e} \lesssim 1$	+1	$\Omega_p < \omega_r <<$ $\min(\lvert\Omega_e\rvert, \omega_e)$
Electron cyclotron	$\omega_r \simeq \lvert\Omega_e\rvert \cos\theta$	$1 < \frac{kc}{\omega_e}$	+ 1	$\lvert\Omega_e\rvert \lesssim \omega_e$
Z mode	$\omega_e < \omega_r < \lvert\Omega_e\rvert$	All k	+1	$\omega_e < \lvert\Omega_e\rvert$
R-X	$\omega_r \geq kc$	All k	+1	$\max(\lvert\Omega_e\rvert, \omega_e) < \omega_r$
L-O	$\omega_r \geq kc$	All k	−1	$\lvert\Omega_e\rvert < \omega_r$

while the cutoff of the left-hand mode lies far below $|\Omega_e|$. As ω_r and the wavenumber increase, both modes approach the properties of the free space light wave.

The left-hand mode is essentially undamped for all wavenumbers, but the right-hand mode becomes cyclotron resonant ($|\zeta_e^{+1}| < 1$) near $\omega_r = |\Omega_e|$ and experiences moderate cyclotron damping in this frequency regime. As with the whistler mode, electron cyclotron damping increases with β_e. Indeed, at sufficiently large values of β_e not illustrated here, the cold plasma approximation fails near $\omega_r = |\Omega_e|$. Thus for a nontrivial range of θ values near 0° the R-X mode goes continuously through the electron cyclotron frequency to become the Z mode at smaller wavenumbers.

At oblique propagation and for the parameters shown in Figure 6.12, the sense of polarization is retained by both the right-hand mode ($0 < \text{Re}(P) \leq 1$) and the left-hand mode ($-1 \leq \text{Re}(P) < 0$). The fluctuating electric fields are, in the perpendicular propagation limit, oriented perpendicular and parallel to $\mathbf{B}_o$ for the two modes, so that once again the notations "R-X" and "L-O" are appropriate.

6.3 Summary

In the low frequency regime, MHD theory predicts the propagation of three normal modes of an isotropic plasma. Vlasov theory also describes these modes but, unlike MHD theory, predicts that some of these modes are heavily damped in certain parameter regimes. Furthermore, among these Vlasov modes, we have identified four distinct waves with quite distinct properties. Table 6.1 summarizes our identification of these waves; Table 6.2 summarizes transport ratios for these waves. Table 6.3 summarizes the electromagnetic normal modes over all frequencies that have been discussed in this Chapter.

Using fluid theories to determine wave properties, several authors (Lacombe *et al.*, 1990; Gleaves and Southwood, 1991) claim to have identified the MHD slow mode in the $T_e < T_p$ magnetosheath. However, Figure 6.1 shows that, if $T_e \lesssim T_p$, the low frequency mode of slowest phase speed is strongly damped at low β_p and is equivalent to the Alfvén wave at high β_p. Furthermore, Figure 6.2 also shows that the Alfvén wave exhibits negative parallel compressibility at sufficiently oblique propagation. Because $C_{\parallel p} < 0$ corresponds to an anticorrelation between the fluctuating density and $\delta B_{\parallel}$, and because this anticorrelation is frequently used as an identifier for the MHD slow mode, we are not convinced that this mode has, in fact, been identified in the magnetosheath. Rather, we believe that the magnetosheath observations indicate the presence of the mirror instability which, as we discuss in Chapter 7, has dispersion and transport properties quite different from those of the wave of slowest phase speed.

References

Barnes, A., Collisionless damping of hydromagnetic waves, *Phys. Fluids*, **9**, 1483, 1966.

Boyd, T. J. M., and J. J. Sanderson, *Plasma Dynamics*, Barnes & Noble, New York, 1969.

Gary, S. P., Low-frequency waves in a high-beta collisionless plasma: polarization, compressibility and helicity, *J. Plasma Phys.*, **35**, 431, 1986.

Gary, S. P., The mirror and ion cyclotron anisotropy instabilities, *J. Geophys. Res.*, **97**, 8519, 1992.

Gary, S. P., and M. M. Mellott, Whistler damping at oblique propagation: laminar shock precursors, *J. Geophys. Res.*, **90**, 99, 1985.

Gary, S. P., and D. Winske, Correlation function ratios and the identification of space plasma instabilities, *J. Geophys. Res.*, **97**, 3103, 1992.

Gleaves, D. G., and D. J. Southwood, Magnetohydrodynamic fluctuations in the Earth's magnetosheath at 1500 LT: ISEE 1 and ISEE 2, *J. Geophys. Res.*, **96**, 129, 1991.

Hada, T., and C. F. Kennel, Nonlinear evolution of slow waves in the solar wind, *J. Geophys. Res.*, **90**, 531, 1985.

Lacombe, C., E. Kinzelin, C. C. Harvey, D. Hubert, A. Mangeney, J. Elaoufir, D. Burgess, and C. T. Russell, Nature of the turbulence observed by ISEE 1-2 during a quasi-perpendicular crossing of the Earth's bow shock, *Ann. Geophys.*, **8**, 489, 1990.

Tokar, R. L., and S. P. Gary, The whistler mode in a Vlasov plasma, *Phys. Fluids*, **28**, 1063, 1985.

7
Electromagnetic temperature anisotropy instabilities in uniform plasmas

In this chapter we examine, as before, electromagnetic fluctuations in a homogeneous, magnetized, collisionless plasma. In contrast to the previous chapter, however, we admit anisotropies in the distribution functions. In particular we consider a two-temperature bi-Maxwellian zeroth-order distribution function; this permits the growth of temperature anisotropy instabilities. Section 7.1 outlines the derivation of the dispersion equation; Section 7.2 discusses the properties of modes driven unstable by a proton temperature anisotropy, whereas Section 7.3 discusses the properties of electron temperature anisotropy instabilities. Section 7.4 is a brief summary.

Our emphasis in this chapter is on instabilities driven by $T_\perp > T_\parallel$, a condition that is observed more often in space plasmas than the converse $T_\parallel > T_\perp$. The reason for this discrepancy is simple: although space plasmas do not necessarily exhibit a bias toward perpendicular heating processes, perpendicular heating does not much change the mobility of the heated particles, whereas parallel heating enables the particles to move more rapidly along $\mathbf{B}_o$. Thus parallel-heated particles may leave the region of energization more quickly, implying that $T_\parallel > T_\perp$ should be a less frequently observed condition. Of course, parallel-heated particles may appear elsewhere as a magnetic-field aligned beam streaming against a cooler background plasma; the electromagnetic instabilities driven by such configurations have quite different properties from temperature anisotropy instabilities, and are studied in detail in the next chapter.

Although many plasma processes such as magnetic compression or expansion lead to temperature anisotropies in both the ion and electron distributions, we here consider instabilities driven by only one species anisotropy at a time. The arguments justifying this approach are different for the two species. For the electrons, the whistler anisotropy instability typically grows at relatively fast times that scale as the electron cyclotron frequency; the ions are not significant participants in this mode and the value of $T_{\perp i}/T_{\parallel i}$ does not affect the properties of this instability. In contrast, electrons with $T_{\perp e}/T_{\parallel e} > 1$ can contribute to the mirror instability, but, because of scattering by the much faster whistler instability, are likely to become relatively isotropic before the ion anisotropy instabilities exhibit much growth.

7.1 The electromagnetic dispersion equation

7.1.1 $\mathbf{k} \times \mathbf{B}_o \neq 0$

The derivation of the dispersion equation at oblique propagation follows the procedure in Section 5.1 through Equation (5.1.8). If we consider a bi-Maxwellian zeroth-order distribution

$$f_j^{(0)}(v_z, v_\perp) = \frac{n_j}{(2\pi v_j^2)^{3/2}} \frac{T_{\parallel j}}{T_{\perp j}} \exp\left[-\frac{v_z^2}{2v_j^2} - \frac{v_x^2 + v_y^2}{2v_j^2}\frac{T_{\parallel j}}{T_{\perp j}}\right], \tag{7.1.1}$$

the first-order distribution follows from Equation (5.1.8):

$$\begin{aligned} f_j^{(1)}(\mathbf{k}, \mathbf{v}, \omega) = {} & \frac{e_j}{T_j} f_j^{(0)}(v_z, v_\perp) \int_{-\infty}^{0} d\tau\, \mathbf{v}' \cdot \mathbf{E}^{(1)}(\mathbf{k}, \omega) \exp[ib_j(\tau, \omega)] \\ & -\frac{e_j}{T_j}\left(1 - \frac{T_{\parallel j}}{T_{\perp j}}\right) f_j^{(0)}(v_z, v_\perp) \\ & \times \int_{-\infty}^{0} d\tau \left[\mathbf{v}'_\perp\left(1 - \frac{k_z v_z}{\omega}\right) + \frac{\hat{\mathbf{z}} k_y v'_y v_z}{\omega}\right] \cdot \mathbf{E}^{(1)}(\mathbf{k}, \omega) \exp[ib_j(\tau, \omega)]. \end{aligned} \tag{7.1.2}$$

The dimensionless conductivity tensor is then, using Equation (5.1.1),

$$\begin{aligned} \mathbf{S}_j(\mathbf{k}, \omega) = {} & \frac{ik_j^2 \omega}{k^2 c^2 n_j} \int d^3v\, \mathbf{v} f_j^{(0)}(v_z, v_\perp) \int_{-\infty}^{0} d\tau\, \mathbf{v}' \exp[ib_j(\tau, \omega)] \\ & -\frac{ik_j^2 \omega}{k^2 c^2 n_j}\left(1 - \frac{T_{\parallel j}}{T_{\perp j}}\right) \int d^3v\, \mathbf{v} f_j^{(0)}(v_z, v_\perp) \\ & \times \int_{-\infty}^{0} d\tau \left[\mathbf{v}'_\perp\left(1 - \frac{k_z v_z}{\omega}\right) + \frac{\hat{\mathbf{z}} k_y v'_y v_z}{\omega}\right] \exp[ib_j(\tau, \omega)]. \end{aligned} \tag{7.1.3}$$

Evaluating this by the first method outlined in Appendix C, we obtain

$$S_{xxj}(\mathbf{k}, \omega) = \frac{\omega_j^2}{k^2 c^2} \zeta_j^0 \exp(-\lambda_j) \sum_{m=-\infty}^{\infty} \left\{2\lambda_j [I_m(\lambda_j) - I'_m(\lambda_j)] + \frac{m^2}{\lambda_j} I_m(\lambda_j)\right\} Z(\zeta_j^m)$$

$$-\frac{\omega_j^2}{2k^2c^2}\left(\frac{T_{\perp j}}{T_{\parallel j}}-1\right)\exp(-\lambda_j)\sum_{m=-\infty}^{\infty}\left\{2\lambda_j[I_m(\lambda_j)-I'_m(\lambda_j)]+\frac{m^2}{\lambda_j}I_m(\lambda_j)\right\}Z'(\zeta_j^m)$$

$$S_{xyj}(\mathbf{k},\omega)=i\frac{\omega_j^2}{k^2c^2}\zeta_j^0\exp(-\lambda_j)\sum_{m=-\infty}^{\infty}m[I_m(\lambda_j)-I'_m(\lambda_j)]Z(\zeta_j^m)$$

$$-i\frac{\omega_j^2}{2k^2c^2}\left(\frac{T_{\perp j}}{T_{\parallel j}}-1\right)\exp(-\lambda_j)\sum_{m=-\infty}^{\infty}m[I_m(\lambda_j)-I'_m(\lambda_j)]Z'(\zeta_j^m)$$

$$S_{xzj}(\mathbf{k},\omega)=i\frac{\omega_j^2}{k^2c^2}\zeta_j^0\frac{|k_z|}{\sqrt{2}k_z}\frac{k_yv_j}{\Omega_j}\exp(-\lambda_j)\sum_{m=-\infty}^{\infty}[I_m(\lambda_j)-I'_m(\lambda_j)]Z'(\zeta_j^m)$$

$$+i\frac{\omega_j^2}{k^2c^2}\left(\frac{T_{\perp j}}{T_{\parallel j}}-1\right)\frac{|k_z|}{\sqrt{2}k_z}\frac{k_yv_j}{\Omega_j}\exp(-\lambda_j)\sum_{m=-\infty}^{\infty}[I_m(\lambda_j)-I'_m(\lambda_j)]\zeta_j^mZ'(\zeta_j^m)$$

$$S_{yxj}(\mathbf{k},\omega)=-S_{xyj}(\mathbf{k},\omega) \tag{7.1.4}$$

$$S_{yyj}(\mathbf{k},\omega)=\frac{\omega_j^2}{k^2c^2}\zeta_j^0\frac{\exp(-\lambda_j)}{\lambda_j}\sum_{m=-\infty}^{\infty}m^2I_m(\lambda_j)Z(\zeta_j^m)$$

$$-\frac{\omega_j^2}{2k^2c^2}\left(\frac{T_{\perp j}}{T_{\parallel j}}-1\right)\frac{\exp(-\lambda_j)}{\lambda_j}\sum_{m=-\infty}^{\infty}m^2I_m(\lambda_j)Z'(\zeta_j^m)$$

$$S_{yzj}(\mathbf{k},\omega)=\frac{\omega_j^2}{k^2c^2}\zeta_j^0\left(\frac{T_{\perp j}}{T_{\parallel j}}\right)\frac{|k_z|}{\sqrt{2}k_z}\frac{k_yv_j}{\Omega_j}\frac{\exp(-\lambda_j)}{\lambda_j}\sum_{m=-\infty}^{\infty}mI_m(\lambda_j)Z'(\zeta_j^m)$$

$$+\frac{\omega_j^2}{2k^2c^2}\left(\frac{T_{\perp j}}{T_{\parallel j}}-1\right)\frac{k_y}{k_z}\frac{\exp(-\lambda_j)}{\lambda_j}\sum_{m=-\infty}^{\infty}m^2I_m(\lambda_j)Z'(\zeta_j^m)$$

$$S_{zxj}(\mathbf{k},\omega)=-S_{xzj}(\mathbf{k},\omega)$$

$$S_{zyj}(\mathbf{k},\omega)=S_{yzj}(\mathbf{k},\omega)$$

$$S_{zzj}(\mathbf{k},\omega)=-\frac{\omega_j^2}{k^2c^2}\zeta_j^0\exp(-\lambda_j)\sum_{m=-\infty}^{\infty}I_m(\lambda_j)\zeta_j^mZ'(\zeta_j^m)$$

$$-\frac{\omega_j^2}{k^2c^2}\left(1-\frac{T_{\parallel j}}{T_{\perp j}}\right)\frac{\Omega_j}{\sqrt{2}|k_z|v_j}\exp(-\lambda_j)\sum_{m=-\infty}^{\infty}mI_m(\lambda_j)\zeta_j^mZ'(\zeta_j^m)$$

where $\lambda_j \equiv (k_y v_j/\Omega_j)^2(T_{\perp j}/T_{\parallel j})$, $Z(\zeta_j^m)$ is the plasma dispersion function (Appendix A), and

$$\zeta_j^m \equiv \frac{\omega + m\Omega_j}{\sqrt{2}|k_z|v_j}.$$

7.1.2 $\mathbf{k} \times \mathbf{B}_o = 0$

At $\mathbf{k} \times \mathbf{B}_o = 0$, there is $k_y = 0$. Writing $k_z^2 = k^2$, the above expressions reduce to

$$S_{xxj}(\mathbf{k}, \omega) = \frac{\omega_j^2}{2k^2c^2}\left\{\zeta_j^0\left[Z(\zeta_j^{+1}) + Z(\zeta_j^{-1})\right] - \frac{1}{2}\left(\frac{T_{\perp j}}{T_{\parallel j}} - 1\right)[Z'(\zeta_j^{+1}) + Z'(\zeta_j^{-1})]\right\}$$

$$S_{xyj}(\mathbf{k}, \omega) = -\frac{i}{2}\frac{\omega_j^2}{k^2c^2}\left\{\zeta_j^0\left[Z(\zeta_j^{+1}) - Z(\zeta_j^{-1})\right] - \frac{1}{2}\left(\frac{T_{\perp j}}{T_{\parallel j}} - 1\right)[Z'(\zeta_j^{+1}) - Z'(\zeta_j^{-1})]\right\}$$

$$S_{xzj}(\mathbf{k}, \omega) = 0$$

$$S_{yxj}(\mathbf{k}, \omega) = -S_{xyj}(\mathbf{k}, \omega)$$

$$S_{yyj}(\mathbf{k}, \omega) = S_{xxj}(\mathbf{k}, \omega)$$

$$S_{yzj}(\mathbf{k}, \omega) = 0$$

$$S_{zxj}(\mathbf{k}, \omega) = 0$$

$$S_{zyj}(\mathbf{k}, \omega) = 0$$

$$S_{zzj}(\mathbf{k}, \omega) = -\frac{k_j^2}{k^2}\frac{v_j^2}{c^2}(\zeta_j^0)^2 Z'(\zeta_j^0).$$

Thus, at $\mathbf{k} \times \mathbf{B}_o = 0$, Equation (5.1.2) factors into two parts. For electrostatic modes with $E_z^{(1)} \neq 0$,

$$D_{zz} = \omega^2\left[1 - \sum_j \frac{k_j^2}{2k^2} Z'\left(\frac{\omega}{\sqrt{2}|k_z|v_j}\right)\right] = 0 \tag{7.1.5}$$

which, at $\omega^2 \neq 0$, yields the electrostatic dispersion equation for Maxwellian distributions at $k_y = 0$.

From the remainder of the dispersion equation one obtains Equation (5.1.10) from which follows the electromagnetic dispersion equation for

$\mathbf{k} \times \mathbf{B}_o = 0$:

$$\omega^2 - k^2c^2 + k^2c^2 \sum_j S_j^{\pm}(\mathbf{k}, \omega) = 0 \tag{7.1.6}$$

where

$$S_j^{\pm}(\mathbf{k}, \omega) = \frac{\omega_j^2}{k^2c^2}\left[\zeta_j^0 Z(\zeta_j^{\pm 1}) - \left(\frac{T_{\perp j}}{T_{\parallel j}} - 1\right)\frac{Z'(\zeta_j^{\pm 1})}{2}\right]. \tag{7.1.7}$$

If $|\gamma| \ll |\omega_r|$, and if we insert bi-Maxwellian distributions into Equation (5.1.16), then for $k_z > 0$

$$\gamma = \frac{\pi}{2\omega_r}\sum_j \frac{\omega_j^2}{(2\pi v_j^2)^{1/2}}\left[-\frac{T_{\perp j}}{T_{\parallel j}}\frac{\omega_r}{k_z} \pm \left(1 - \frac{T_{\perp j}}{T_{\parallel j}}\right)\frac{\Omega_j}{k_z}\right]\exp\left[\frac{(\omega_r \pm \Omega_j)^2}{2k_z^2 v_j^2}\right]. \tag{7.1.8}$$

From this expression and Equation (5.2.1) it is clear that, for modes with $0 < \omega_r \ll |\Omega_j|$, an ion (electron) species with $T_{\perp j} < T_{\parallel j}$ has the potential to drive a right- (left-) hand polarized mode unstable, whereas $T_{\perp j} > T_{\parallel j}$ corresponds to a left- (right-) hand instability.

Problem 7.1.1. Consider an electromagnetic wave propagating parallel to $\mathbf{B}_o$ with $\omega_r/k_z > 0$ in the plasma frame. If a gyrating ion is to experience a cyclotron resonance with this wave, find its average field-aligned velocity and the wave helicity if the wave is (a) right-hand polarized and (b) left-hand polarized in the plasma frame.

7.2 Electromagnetic ion temperature anisotropy instabilities

In this section, we consider a two-species electron-proton plasma in which the electrons are represented by a single Maxwellian distribution and the protons have a bi-Maxwellian distribution. Because the numerical examples here use $m_i = m_p = 1836m_e$, we use the term "proton", rather than "ion" throughout this section; the distinction between protons and heavier ions as the source of free energy is important, for example, in discussing cyclotron anisotropy instabilities.

7.2.1 The proton firehose instability

If $T_{\parallel p} > T_{\perp p}$, the proton firehose instability may arise in a plasma with sufficiently large β_p. Kennel and Petschek (1966) were the first to point out that ions resonate with this instability and that $\omega_r \sim \Omega_p$. The dispersion properties of this mode are illustrated in Figure 7.1. The instability has

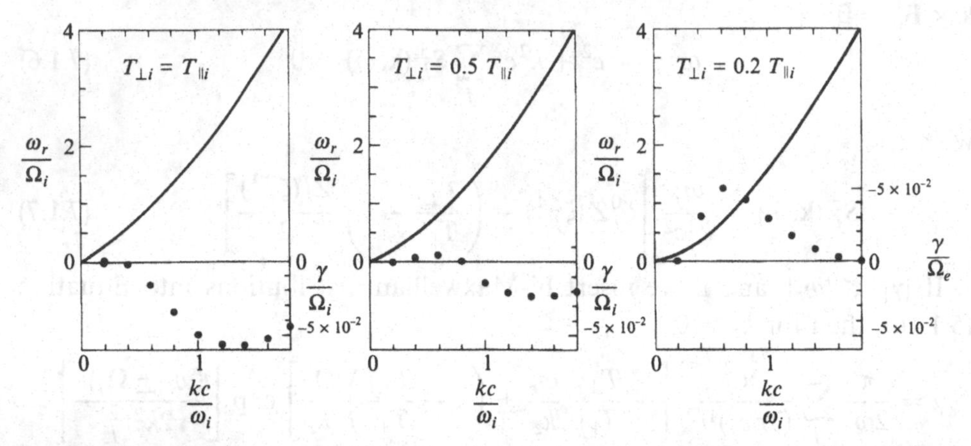

Fig. 7.1 The real frequency (solid lines) and damping/growth rate (dotted lines) of the magnetosonic/whistler mode and proton firehose instability as functions of wavenumber at propagation parallel to $\mathbf{B}_o$ for three different values of $T_{\perp p}/T_{\parallel p}$. Here, and for all the figures of this chapter except as noted, we consider an electron-proton plasma with $m_i = m_p = 1836m_e$, $T_e = T_p$ and $v_A/c = 10^{-4}$. Here $\beta_p = 2.0$.

maximum growth at $\mathbf{k} \times \mathbf{B}_o = 0$ (Gary *et al.*, 1976), is right-hand circularly polarized in this direction of propagation, and evolves out of the magnetosonic/whistler wave as the anisotropy increases. For this instability the electrons are nonresonant ($|\zeta_e^{+1}| \gg 1$), but the protons are cyclotron resonant ($|\zeta_p^{+1}| > 1$) (Fig. 4 of Gary and Feldman, 1978). Because the electrons are nonresonant, the growth rate is essentially independent of T_e/T_p.

7.2.2 The proton cyclotron anisotropy instability

If $T_{\perp p} > T_{\parallel p}$, several different instabilities may arise (Gary *et al.*, 1976). Under the conditions $\beta_p \lesssim 6$ and $T_e \sim T_p$ in an electron-proton plasma, the fastest-growing mode is the proton cyclotron anisotropy instability (Kennel and Petschek, 1966), which has maximum growth at $\mathbf{k} \times \mathbf{B}_o = 0$, has left-hand circular polarization at that direction of propagation, and evolves out of the Alfvén/proton cyclotron wave if the proton anisotropy is gradually increased (Figure 7.2). Representative dispersion plots for this instability are shown in Figure 7.2 and Figs. 2 and 3 of Davidson and Ogden (1975). For this instability the electrons are nonresonant ($|\zeta_e^{-1}| \gg 1$), but the protons are cyclotron resonant ($|\zeta_p^{-1}| \gtrsim 1$) (Fig. 2 of Gary and Feldman, 1978). Because the electrons are nonresonant, the growth rate is essentially independent of T_e/T_p.

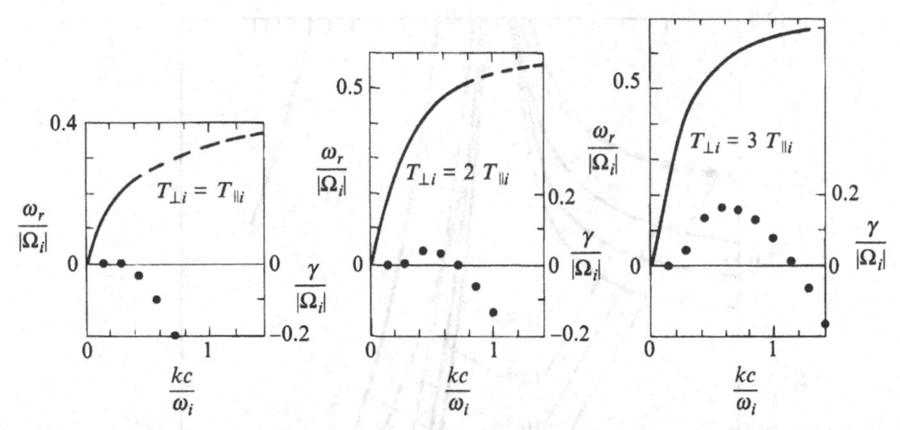

Fig. 7.2 The real frequency (solid/dashed lines) and damping/growth rates (dotted lines) of the Alfvén/proton cyclotron mode and the proton cyclotron anisotropy instability as functions of wavenumber at propagation parallel to $\mathbf{B}_o$ for three different values of $T_{\perp p}/T_{\parallel p}$. Here ω_r is indicated by the solid lines in the regime of light damping ($|\gamma| \leq |\omega_r|/2\pi$) and by dashed lines in the heavily damped regime ($\gamma < -|\omega_r|/2\pi$). Here $\beta_p = 1.0$.

The maximum growth rates for the proton firehose instability and the proton cyclotron anisotropy instability as functions of β_p and temperature anisotropy are illustrated in Figure 7.3.

7.2.3 The mirror instability

The mirror instability (Chandrasekhar *et al.*, 1958; Barnes, 1966) may also grow under the condition $T_{\perp p} > T_{\parallel p}$. This mode is a Landau resonant ($|\zeta_p^0| \ll 1$) instability that exists in the long wavelength MHD limit, has $\omega_r = 0$ in a uniform plasma, and exhibits maximum growth rate at wavevector directions oblique to $\mathbf{B}_o$. Thus, unlike the predominantly transverse fluctuations of the proton firehose and proton cyclotron anisotropy instabilities, the fluctuating magnetic fields of the mirror instability have a substantial longitudinal component; that is, $\delta\mathbf{B}$ has a significant component parallel to $\mathbf{B}_o$. The growth rate as a function of wavenumber for three different values of the temperature anisotropy is illustrated in Figure 7.4. Note that an increase in $T_{\perp p}/T_{\parallel p}$ corresponds not only to an increase in the maximum growth rate, but also to a shift of the direction of propagation of γ_m toward directions more nearly parallel to $\mathbf{B}_o$.

Gary *et al.* (1976) used linear theory to show that, in an electron-proton plasma with $T_e = T_p$ and bi-Maxwellian distributions, the proton cyclotron anisotropy instability has a lower anisotropy threshold than the mirror in-

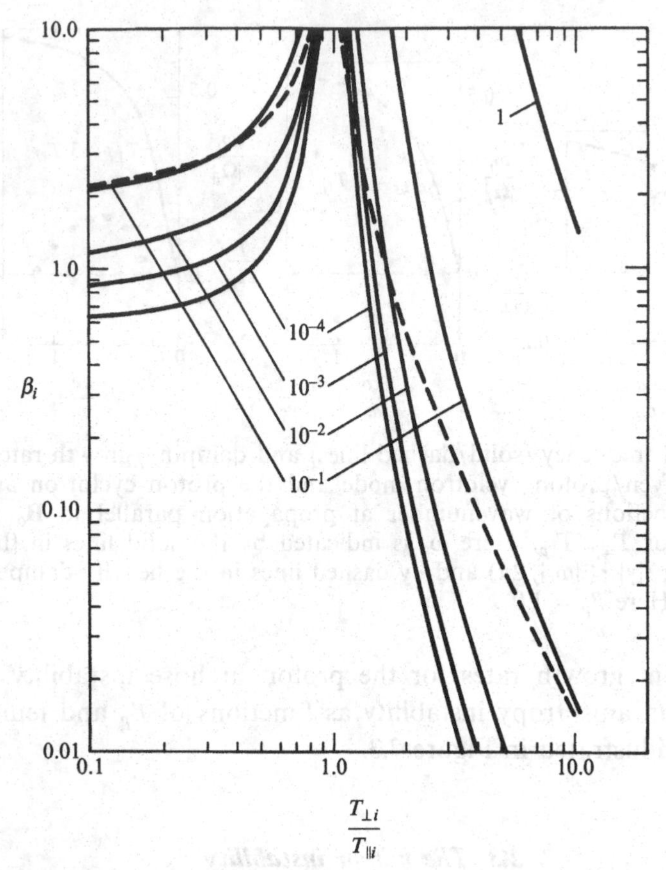

Fig. 7.3 Curves of constant maximum growth rate γ_m/Ω_p in the $(\beta_p, T_{\perp p}/T_{\parallel p})$-plane. The solid lines represent the proton firehose $(T_{\parallel p} > T_{\perp p})$ and the proton cyclotron anisotropy $(T_{\perp p} > T_{\parallel p})$ instabilities; the dashed lines represent the MHD thresholds corresponding to the nonresonant firehose $(T_{\parallel p} > T_{\perp p})$ and the mirror $(T_{\perp p} > T_{\parallel p})$ instabilities. Here $v_A/c = 1.4 \times 10^{-4}$ (From Fig. 3 of Gary *et al.*, 1976).

stability at $\beta_p \lesssim 6.0$ (Figure 7.3). More recently, Gary (1992) has shown that in an electron-proton plasma with $T_e = T_p$ the proton cyclotron anisotropy instability has the larger growth rate for a wide range of conditions. Low-β conditions under which the mirror may have the larger γ_m include $T_{\perp e}/T_{\parallel e} \gg 1$ or, if the anisotropy is not too far above threshold, the presence of a small amount of helium ions (Price *et al.*, 1986; Gary, 1992).

7.2.4 Transport ratios for $T_{\perp p} > T_{\parallel p}$ instabilities

Figure 7.5 compares the growth rate and five transport ratios of the mirror and proton cyclotron anisotropy instabilities as functions of θ at $kc/\omega_p =$

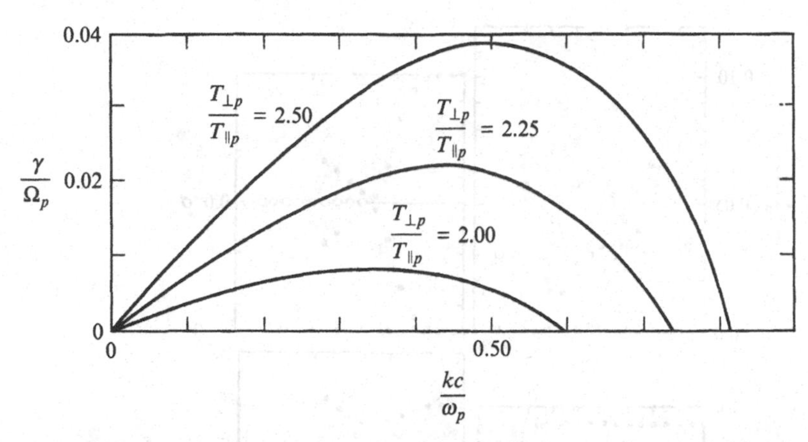

Fig. 7.4 The growth rate of the mirror instability as a function of wavenumber at three different values of the proton temperature anisotropy. Here $\beta_p = 1.00$. For each curve, the angle of propagation corresponds to the maximum growth rate at that value of the anisotropy: at $T_{\perp p}/T_{\parallel p} = 2.00$, $\theta = 71°$; at $T_{\perp p}/T_{\parallel p} = 2.25$, $\theta = 66°$; at $T_{\perp p}/T_{\parallel p} = 2.50$, $\theta = 63°$ (From Fig. 2 of Gary, 1992).

0.375, which is approximately the wavenumber of maximum growth for both instabilities under the parameters used here. The proton cyclotron instability has maximum growth at parallel propagation; the "shoulder" in $\gamma(\theta)$ corresponds to the Harris instability; and the mirror mode is unstable only at oblique propagation angles.

Because the proton cyclotron anisotropy instability propagates symmetrically with respect to $\mathbf{B}_o$, and the helicity and cross-helicity change sign under a reversal of propagation direction, we have plotted two values of σ and σ_{cp} for this mode; the negative values of these quantities at small θ correspond to $\omega_r/k_z > 0$. Note that both the helicity and the cross-helicity of this instability at nonzero wavenumber have the same qualitative θ dependence as they do for the Alfvén/proton cyclotron mode in an isotropic plasma (Gary, 1986). For the mirror mode, both σ and σ_{cp} are identically zero. Although the two instabilities have different magnitudes for these two transport ratios, the possibility that diagnostics might not be able to resolve right- from left-travelling fluctuations of the proton cyclotron mode suggests that neither the helicity nor the cross-helicity may be useful in distinguishing the two instabilities. The proton compressibility shows similar values at oblique propagation of the two instabilities, indicating that it, too, is probably not a useful identifier in this case. And although σ_{EE} of the proton cyclotron anisotropy instability is consistently larger than σ_{EE} of the mirror instability, both quantities are typically orders of magnitude smaller than unity, and would probably not be helpful in discerning between the two modes.

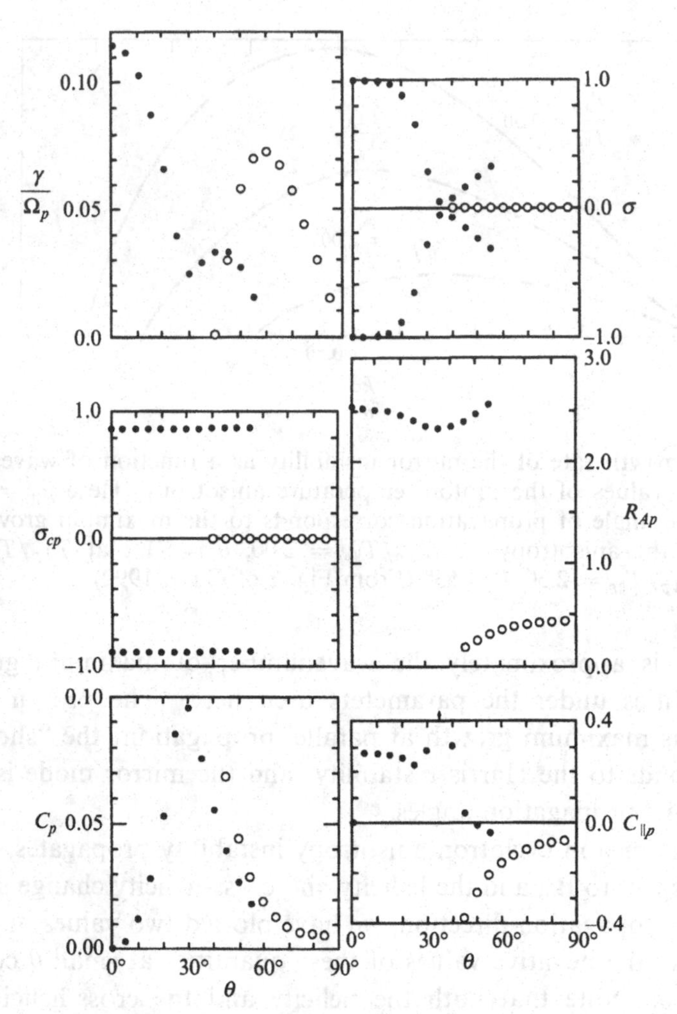

Fig. 7.5 The growth rate and five transport ratios of the proton cyclotron anisotropy instability (solid dots) and the mirror instability (open dots) as functions of propagation angle. Here $\beta_p = 4.0$, $T_{\perp p}/T_{\parallel p} = 2.0$ and $kc/\omega_p = 0.375$. The proton cyclotron anisotropy instability at $\omega_r/k_z > 0$ corresponds to $\sigma < 0$ at $\theta \lesssim 35°$ and $\sigma_{cp} < 0$ for all θ (From Fig. 7 of Gary, 1992).

In contrast, Figure 7.5 shows that both the parallel compressibility and the Alfvén ratio are quite different for the two instabilities and can be used to distinguish between them. The negative values of $C_{\parallel p}$ for the mirror instability reflect the well-known MHD property that density and B_z fluctuations are anticorrelated; the positive values of the parallel compressibility for the proton cyclotron instability indicates the correlation of these two quantities for this mode. Although the Alfvén/proton cyclotron mode in an isotropic plasma typically satisfies $R_{Ap} \simeq 1$ (Fig. 6 of Gary, 1986), the phase speed of the proton cyclotron instability increases as the temperature anisotropy increases (Fig. 3 of Davidson and Ogden, 1975), and the Alfvén ratio becomes substantially greater than unity. In contrast, $R_{Ap} \ll 1$ for the mirror instability. The computer simulations of McKean *et al.* (1992b) have demonstrated that at relatively weak fluctuation levels linear theory provides a good approximation to the transport ratios for these modes, and that the Alfvén ratio in particular can be useful in distinguishing between the two instabilities.

Problem 7.2.1. Show that, if $\omega_r = 0$, the helicity is zero. (Hint: if you can prove that all the elements of $\mathbf{S}_j(\mathbf{k}, \omega)$ are real, then Equations (5.1.5) and (5.1.6) provide useful information.)

7.2.5 Applications of $T_{\perp p} > T_{\parallel p}$ instabilities

Electromagnetic instabilities due to $T_{\perp p} > T_{\parallel p}$ are often invoked in studies of the magnetosphere, where such anisotropies are often observed and the plasma β values are large enough to sustain appreciable growth rates. The presence of distinct hot and cold plasma components in the magnetosphere can, in particular, be conducive to growth of these instabilities. Cornwall *et al.* (1970) and Cornwall and Schulz (1971) showed that the presence of a cold ion component could enhance the proton cyclotron anisotropy instability driven by a hot proton component with $T_{\perp p}/T_{\parallel p} > 1$. Reviews of theoretical and computer simulation research on temperature anisotropy instabilities in the magnetosphere include Cuperman (1981) and Gendrin *et al.* (1984).

The magnetosheath is a good example of a space plasma regime in which the observed condition $T_{\perp p} > T_{\parallel p}$ has been correlated with the observation of enhanced low-frequency magnetic fluctuations. The heating and reflection of solar wind ions at the Earth's bow shock produces strongly anisotropic ion distribution functions immediately downstream of that shock. As the plasma flows into the magnetosheath, these distributions often assume bi-Maxwellian forms such that $T_{\perp p}/T_{\parallel p} > 1$ (Thomsen *et al.*, 1985; Sckopke *et al.*, 1990; Anderson *et al.*, 1991). Still deeper in the magnetosheath, a region of reduced plasma density, enhanced B_o, and reduced plasma β often forms just upsteam from the magnetopause (Crooker *et al.*, 1979; Song *et al.*, 1990). Anderson *et al.* (1991), Fuselier *et al.* (1991), and Anderson and Fuselier (1993), have named this region the plasma depletion layer and have noted that the proton temperature anisotropy is larger there than in the magnetosheath proper.

Because the sheath is typically a high β_p regime, both the proton cyclotron anisotropy instability and the mirror instability have the potential to grow and produce the observed reduction in the proton temperature anisotropy. The mirror mode has been identified in the magnetosheath by many authors primarily through the anticorrelation of the magnetic field and density fluctuations (Kaufmann *et al.*, 1970; Crooker *et al.*, 1979; Tsurutani *et al.*, 1982; Moustaizis *et al.*, 1986; Hubert *et al.*, 1989a, 1989b). Sckopke *et al.* (1990) report observations of enhanced proton-cyclotron-like fluctuations in the sheath. Recently, Anderson *et al.* (1991), Fuselier *et al.* (1991), and Anderson and Fuselier (1993) have shown that proton-cyclotron-like fluctuations arise preferentially in the plasma depletion layer, whereas mirror-like fluctuations are more likely to be observed in the magnetosheath proper. Because the cyclotron fluctuation observations typically correspond to the $\beta \ll 1$, high-anisotropy conditions of low-β, low Mach shocks and the plasma depletion layer, and the mirror observations correspond to the $\beta \geq 1$, lower anisotropy conditions of the magnetosheath proper, it is clear that β_p and $T_{\perp p}/T_{\parallel p}$ play important roles in determining which instability will have the larger growth rate and which will dominate the observed fluctuations in a given magnetosheath regime. However, these two parameters do not tell the whole story.

Price *et al.* (1986) showed that the presence of a sufficiently dense He^{++} component reduces the maximum growth rate of the proton cyclotron anisotropy instability so that if the proton anisotropy is not too large, the mirror instability can have the larger linear growth rate at β_p values somewhat less than 6. By including in their linear theory calculations the He^{++} observed by Anderson and Fuselier (1993), Gary *et al.* (1993) have shown that it is the helium component that allows the mirror to dominate wave growth in the magnetosheath proper when $\beta_p \gtrsim 1$ and the proton anisotropy is relatively modest. In contrast, for the larger proton anisotropies of the low-Mach, low-β quasiperpendicular shocks and the plasma depletion layer, the cyclotron instability is well above threshold and, as Gary *et al.* (1993) have shown, can dominate fluctuation growth despite the continued presence of the helium ions.

However, the competition between the mirror instability and the proton cyclotron instability has not yet been critically examined in the fully nonlinear context of computer simulations. Most simulations published to date for both the mirror instability (Price *et al.*, 1986; Lee *et al.*, 1988; McKean *et al.*, 1992a, 1992b) and the proton cyclotron anistropy instability (e.g. Cuperman, 1981; Tanaka, 1985; Omura *et al.*, 1985; Machida *et al.*, 1988) have been one-dimensional and have not permitted a true competition between the two instabilities. The two-dimensional shock and magnetosheath simulations of Winske and Quest (1988) which utilize a single ion component suggest that the ion cyclotron instability can dominate the sheath fluctuations at both low and high upstream ion beta. Further computer simulations, as well as magnetosheath data analysis to examine the nonlinear competition between these two modes and their response to the presence of ionized helium, would be appropriate.

The magnetosheath spectra of Anderson and Fuselier (1993) also show enhanced transverse magnetic fluctuations at frequencies well below the helium ion cyclotron frequency. Gary *et al.* (1993) have identified these fluctuations as due to the helium cyclotron anisotropy instability, a mode driven unstable by the temperature anisotropy of the tenuous He^{++} component in the magnetosheath. This instability is analogous to the proton cyclotron anisotropy instability in that it also is left-hand polarized and has maximum growth at $\mathbf{k} \times \mathbf{B}_o = 0$, although it has longer wavelengths and lower frequencies than the proton mode. Analytic and computer simulation studies of this instability and its role in magnetosheath dynamics are presently under way.

7.3 Electromagnetic electron temperature anisotropy instabilities

In this section we consider an electron-proton plasma in which the protons are represented by a single Maxwellian distribution, and the electrons have a bi-Maxwellian distribution.

7.3.1 The electron firehose instability

If $T_{\parallel e} > T_{\perp e}$, the electron firehose instability can arise (Hollweg and Volk, 1970; Pilipp and Volk, 1971); a representative dispersion plot for this mode is illustrated in Figure 7.6. This instability evolves out of the right-hand polarized whistler wave at small anisotropy, but undergoes a significant shift in ω_r as $T_{\perp e}/T_{\parallel e}$ decreases so that, in the unstable regime, $\omega_r > 0$ and the mode is left-hand circularly polarized at $\mathbf{k} \times \mathbf{B}_o = 0$. For the electron firehose instability, the electrons are nonresonant ($|\zeta_e^{-1}| \gg 1$), but at $T_e \sim T_p$ in an electron-proton plasma the protons are cyclotron resonant ($|\zeta_p^{-1}| \sim 1$), so that the maximum growth rate of this instability is a weakly increasing function of T_e/T_p (See Fig. 4 of Gary and Madland, 1985).

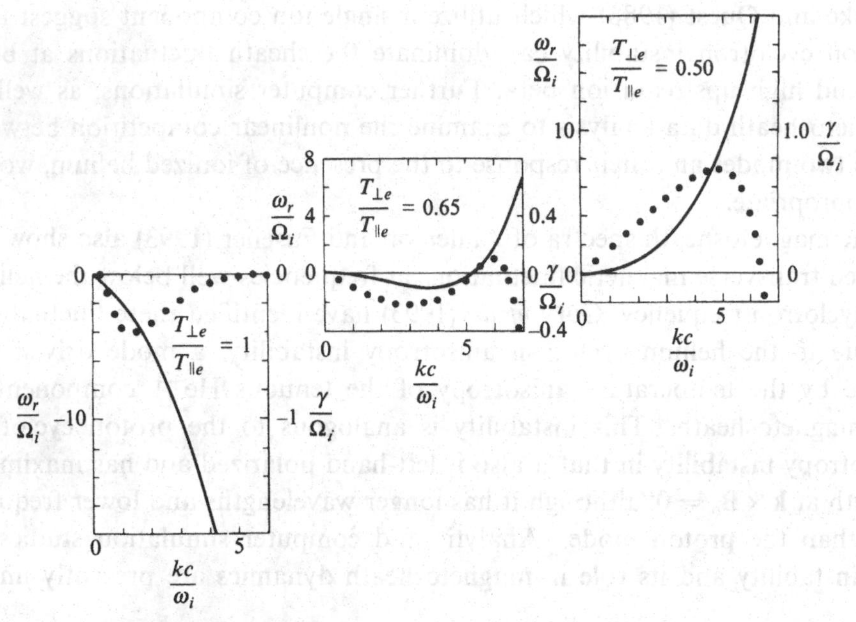

Fig. 7.6 The real frequency (solid lines) and damping/growth rates (dotted lines) for negative helicity fluctuations as functions of wavenumber at $\mathbf{k} \times \mathbf{B}_o = 0$ for three different values of $T_{\perp e}/T_{\parallel e}$ at $\beta_e = 5.0$. The regime of positive γ here corresponds to the electron firehose instability (From Fig. 3 of Gary and Madland, 1985).

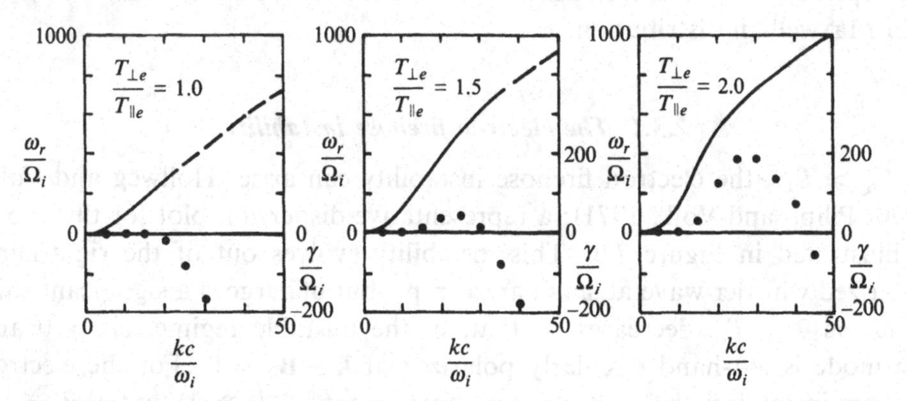

Fig. 7.7 The real frequency (solid/dashed lines) and damping/growth rates (dotted lines) of the whistler mode and the whistler anisotropy instability as functions of wavenumber at $\mathbf{k} \times \mathbf{B}_o = 0$ for three different values of $T_{\perp e}/T_{\parallel e}$ at $\beta_e = 1.0$. Here ω_r is indicated by the solid lines in the regime of light damping ($|\gamma| \leq |\omega_r|/2\pi$) and by dashed lines under heavy damping ($\gamma < -|\omega_r|/2\pi$). Here $\omega_e/|\Omega_e| = 100.0$ (From Fig. 1 of Gary and Madland, 1985).

7.3.2 The whistler anisotropy instability

The opposite case, $T_{\perp e}/T_{\parallel e} > 1$, leads to the whistler anisotropy instability (Kennel and Petschek, 1966; Scharer and Trivelpiece, 1967) illustrated in Figure 7.7. The maximum growth rate of this instability lies at $\mathbf{k} \times \mathbf{B}_o = 0$. As shown in Figure 7.7, an increasing electron anisotropy corresponds to an increase in ω_r at a fixed value of kc/ω_p. The electrons are cyclotron resonant with this instability ($|\zeta_e^{+1}| > 1$), but because the protons are nonresonant ($|\zeta_p^{+1}| \gg 1$), T_e/T_p has no effect on the properties of this instability. Cuperman (1981) reviewed the theoretical and computer simulation literature concerning the whistler anisotropy instability.

Problem 7.3.1. By ignoring the protons and assuming that the electrons are nonresonant $|\zeta_e^{+1}| \gg 1$, show that the dispersion relation for the whistler anisotropy instability at parallel propagation is

$$\omega_r \simeq k^2 c^2 \frac{\Omega_p}{\omega_p^2}\left[1 + \left(\frac{T_{\perp e}}{T_{\parallel e}} - 1\right)\frac{\beta_e}{2}\right].$$

Use Equation (7.1.8) to demonstrate that a necessary condition for this instability is (Kennel and Petscheck, 1966)

$$\frac{T_{\perp e}}{T_{\parallel e}} - 1 > \frac{1}{|\Omega_e|/\omega_r - 1}.$$

Note that both unstable panels of Figure 7.7 satisfy this latter condition.

The maximum growth rates of these two instabilities as functions of β_e and temperature anisotropy are illustrated in Figure 7.8. Note the similarities between Figures 7.3 and 7.8; in both cases the $T_\perp > T_\parallel$ instabilities can persist with appreciable growth rates to relatively low values of β, whereas the $T_\parallel > T_\perp$ instabilities are limited to the $\beta \gtrsim 1$ regime for both species. Although both firehose modes as well as the proton cyclotron anisotropy instability are approximately limited to $\gamma_m \lesssim \Omega_p$, the whistler anisotropy instability can grow with a rate which is an appreciable fraction of $|\Omega_e|$.

7.3.3 Applications of $T_{\perp e} > T_{\parallel e}$ instabilities

The whistler anisotropy instability is a likely source of several different magnetospheric fluctuations including plasmaspheric hiss and magnetospheric chorus (Shawhan, 1979). Chorus is observed in the outer magnetosphere, between the plasmapause and the magnetopause; it is generally thought to be be generated by energetic, anisotropic electrons (Isenberg *et al.*, 1982,

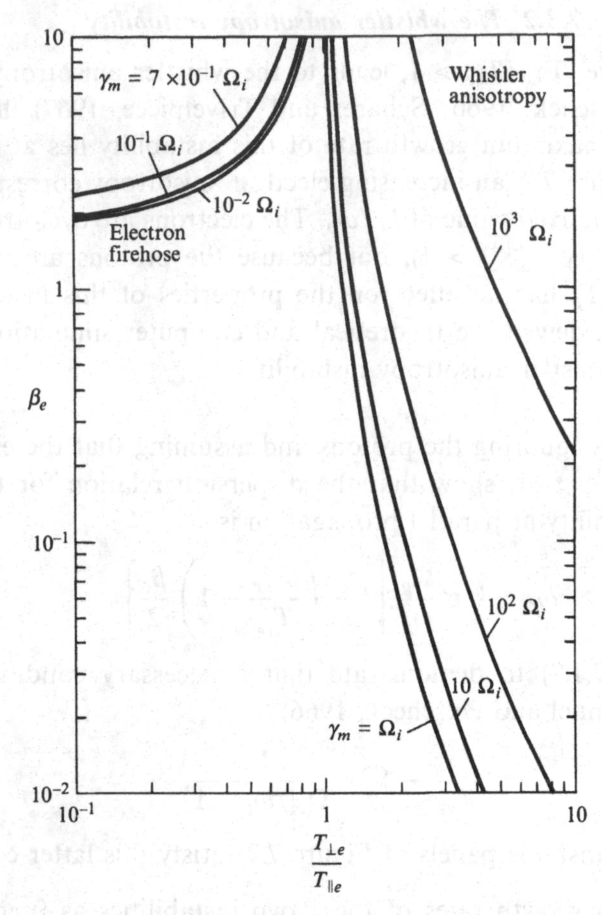

Fig. 7.8 Curves of constant maximum growth rate in the $(\beta_e, T_{\perp e}/T_{\parallel e})$-plane. The solid lines represent the electron firehose $(T_{\parallel e} > T_{\perp e})$ and the whistler anisotropy $(T_{\perp e} > T_{\parallel e})$ instabilities. Here $\omega_e/|\Omega_e| = 100.0$ (From Fig. 5 of Gary and Madland, 1985).

and references therein). However, theoretical analyses indicate that electron loss-cone distributions, rather than simple temperature anisotropies, are necessary to enable detailed agreement between chorus theory and observation (Curtis, 1978).

Further from the Earth, Gosling *et al.* (1989) have reported the observation of $T_{\perp e} > T_{\parallel e}$ in the magnetosheath. Somewhat surprisingly, electrons heated at the quasiperpendicular bow shock are relatively isotropic; the anisotropy in the direction perpendicular to $\mathbf{B}_o$ develops in the suprathermal electrons (speeds greater than about 1000 km/s) with increasing distance into the magnetosheath. These observations have not been correlated with magnetic fluctuation observations, but enhanced whistlers called "lion roars" have been

Table 7.1. *Some electromagnetic temperature anisotropy instabilities*

Name	Source	Frequency	Wavenumber range	γ_m condition and polarization
Proton firehose	$T_{\parallel p} > T_{\perp p}$	$\omega_r << \Omega_p$	$kc/\omega_p \lesssim 1$	$\mathbf{k} \times \mathbf{B}_o = 0$ right-hand
Mirror	$T_{\perp p} > T_{\parallel p}$ $T_{\perp e} > T_{\parallel e}$	$\omega_r = 0$	$kc/\omega_p \lesssim 1$	$\mathbf{k} \times \mathbf{B}_o \neq 0$
Proton cyclotron anisotropy	$T_{\perp p} > T_{\parallel p}$	$0 < \omega_r < \Omega_p$	$kc/\omega_p \lesssim 1$	$\mathbf{k} \times \mathbf{B}_o = 0$ left-hand
Electron firehose	$T_{\parallel e} > T_{\perp e}$	$\Omega_p < \omega_r << \lvert\Omega_e\rvert$	$kc/\omega_p > 1$	$\mathbf{k} \times \mathbf{B}_o = 0$ left-hand
Whistler anisotropy	$T_{\perp e} > T_{\parallel e}$	$\Omega_p < \omega_r < \lvert\Omega_e\rvert$	$kc/\omega_e \lesssim 1$	$\mathbf{k} \times \mathbf{B}_o = 0$ right-hand

attributed to the whistler anisotropy instability. As described by Tsurutani *et al.* (1982), $T_{\perp p} > T_{\parallel p}$ conditions in the magnetosheath give rise to the mirror instability, which produces large amplitude, anticorrelated fluctuations in the magnetic field magnitude and plasma density. The regions of field minima correspond to high β that allows the whistler anisotropy instability to grow within these minima, in agreement with the observations. Moreira (1983) has done a linear stability analysis of magnetosheath lion roars.

7.4 Summary

Table 7.1 summarizes some properties of the electromagnetic temperature anisotropy instabilities studied in this chapter. These unstable modes propagate at frequencies below the cyclotron frequency of the anisotropic species, typically (except for the mirror instability) propagate with maximum growth rate at $\mathbf{k} \times \mathbf{B}_o = 0$, and have wavenumbers that scale as species inertial lengths. Typically, increasing β reduces the threshold anisotropy and increases the growth rates of these instabilities. Thus these instabilities are more likely to be found in space plasmas of relatively high β such as the magnetosphere, the magnetosheath and the solar wind.

Because $T_{\parallel j} > T_{\perp j}$ is rarely observed in space plasmas, the firehose instabilities driven by temperature instabilities are not frequently reported in the literature. Some authors have applied the term "firehose" to the nonresonant electromagnetic instability driven by the relative streaming of two relatively cold ion components. However, the properties of this ion/ion nonresonant instability are quite different from those of the proton firehose instability driven by a temperature anisotropy; the ion/ion mode is treated in Chapter 8.

Because both the proton cyclotron anisotropy and mirror instabilities may grow in plasmas with $T_{\perp p} > T_{\parallel p}$, there can be competition between the two modes in anisotropic high-β plasmas such as the magnetosheath. At this writing, this is an area of substantial research interest.

References

Anderson, B. J., and S. A. Fuselier, Magnetic pulsations from 0.1 to 4.0 Hz and associated plasma properties in the Earth's subsolar magnetosheath and plasma depletion layer, *J. Geophys. Res.*, **98**, 1461, 1993.

Anderson, B. J., S. A. Fuselier and D. Murr, Electromagnetic ion cyclotron waves observed in the plasma depletion layer, *Geophys. Res. Lett.*, **18**, 1955, 1991.

Barnes, A., Collisionless damping of hydromagnetic waves, *Phys. Fluids*, **9**, 1483, 1966.

Chandrasekhar, S., A. N. Kaufman and K. M. Watson, The stability of the pinch, *Proc. Roy. Soc. London, Ser. A*, **245**, 435, 1958.

Cornwall, J. M., F. V. Coroniti and R. M. Thorne, Turbulent loss of ring current protons, *J. Geophys. Res.*, **75**, 4699, 1970.

Cornwall, J. M., and M. Schulz, Electromagnetic ion-cyclotron instabilities in multicomponent magnetospheric plasmas, *J. Geophys. Res.*, **76**, 7791, 1971.

Crooker, N. U., T. E. Eastman and G. S. Stiles, Observations of plasma depletion in the magnetosheath at the dayside magnetopause, *J. Geophys. Res.*, **84**, 869, 1979.

Cuperman, S., Electromagnetic kinetic instabilities in multicomponent space plasmas: theoretical predicitions and computer simulation experiments, *Revs. Geophys. Space Phys.*, **19**, 307, 1981.

Curtis, S. A., A theory for chorus generation by energetic electrons during substorms, *J. Geophys. Res.*, **83**, 3841, 1978.

Davidson, R. C., and J. M. Ogden, Electromagnetic ion cyclotron instability driven by ion energy anisotropy in high-beta plasmas, *Phys. Fluids*, **18**, 1045, 1975.

Fuselier, S. A., D. M. Klumpar, E. G. Shelley, B. J. Anderson and A. J. Coates, He^{2+} and H^+ dynamics in the subsolar magnetosheath and plasma depletion layer, *J. Geophys. Res.*, **96**, 21095, 1991.

Gary, S. P., Low-frequency waves in a high-beta collisionless plasma: polarization, compressibility and helicity, *J. Plasma Phys.*, **35**, 431, 1986.

Gary, S. P., The mirror and ion cyclotron anisotropy instabilities, *J. Geophys. Res.*, **97**, 8519, 1992.

Gary, S. P., and W. C. Feldman, A second-order theory for $\mathbf{k} \parallel \mathbf{B}_o$ electromagnetic instabilities, *Phys. Fluids*, **21**, 72, 1978.

Gary, S. P., and C. D. Madland, Electromagnetic electron temperature anisotropy instabilities, *J. Geophys. Res.*, **90**, 7607, 1985.

Gary, S. P., M. D. Montgomery, W. C. Feldman and D. W. Forslund, Proton temperature anisotropy instabilities in the solar wind, *J. Geophys. Res.*, **81**, 1241, 1976.

Gary, S. P., S. A. Fuselier and B. J. Anderson, Ion anisotropy instabilities in the magnetosheath, *J. Geophys. Res.*, **98**, 1481, 1993.

Gendrin, R., M. Ashour-Abdalla, Y. Omura and K. Quest, Linear analysis of ion cyclotron interaction in a multicomponent plasma, *J. Geophys. Res.*, **89**, 9119, 1984.

Gosling, J. T., M. F. Thomsen, S. J. Bame and C. T. Russell, Suprathermal electrons at Earth's bow shock, *J. Geophys. Res.*, **94**, 10011, 1989.

Hollweg, J. V., and H. J. Volk, New plasma instabilities in the solar wind, *J. Geophys. Res.*, **75**, 5297, 1970.

Hubert, D., C. Perche, C. C. Harvey, C. Lacombe and C. T. Russell, Observation of mirror waves downstream of a quasiperpendicular shock, *Geophys. Res. Lett.*, **16**, 159, 1989a.

Hubert, D., C. C. Harvey and C. T. Russell, Observations of magnetohydrodynamic modes in the Earth's magnetosheath at 0600 LT, *J. Geophys. Res.*, **94**, 17305, 1989b.

Isenberg, P. A., H. C. Koons and J. F. Fennell, Simultaneous observations of energetic electrons and dawnside chorus in geosynchronous orbit, *J. Geophys. Res.*, **87**, 1495, 1982.

Kaufmann, R. L., J.-T. Horng and A. Wolfe, Large-amplitude hydromagnetic waves in the inner magnetosheath, *J. Geophys. Res.*, **75**, 4666, 1970.

Kennel, C. F., and H. E. Petschek, Limit on stably trapped particle fluxes, *J. Geophys. Res.*, **71**, 1, 1966.

Lee, L. C., C. P. Price, C. S. Wu and M. E. Mandt, A study of mirror waves generated downstream of a quasiperpendicular shock, *J. Geophys. Res.*, **93**, 247, 1988.

Machida, S., C. K. Goertz and T. Hada, The electromagnetic ion cyclotron instability in the Io torus, *J. Geophys. Res.*, **93**, 7545, 1988.

McKean, M. E., D. Winske and S. P. Gary, Kinetic properties of mirror waves in magnetosheath plasmas, *Geophys. Res. Lett.*, **19**, 1331, 1992a.

McKean, M. E., D. Winske and S. P. Gary, Mirror and ion cyclotron anisotropy instabilities in the magnetosheath, *J. Geophys. Res.*, **97**, 19,421, 1992b.

Moreira, A., Stability analysis of magnetosheath lion roars, *Planet. Space Sci.*, **31**, 1165, 1983.

Moustaizis, S., D. Hubert, A. Mangeney, C. C. Harvey, C. Perche and C. T. Russell, Magnetohydrodynamic turbulence in the Earth magnetosheath, *Ann. Geophys.*, **4A**, 355, 1986.

Omura, Y., M. Ashour-Abdalla, R. Gendrin and K. Quest, Heating of thermal helium in the equatorial magnetosphere: A simulation study, *J. Geophys. Res.*, **90**, 8281, 1985.

Pilipp, W., and H. J. Volk, Analysis of electromagnetic instabilites parallel to the magnetic field, *J. Plasma Phys.*, **6**, 1, 1971.

Price, C. P., D. W. Swift and L.-C. Lee, Numerical simulation of nonoscillatory mirror waves at the Earth's magnetosheath, *J. Geophys. Res.*, **91**, 101, 1986.

Scharer, J. E., and A. W. Trivelpiece, Cyclotron wave instabilities in a plasma, *Phys. Fluids*, **10**, 591, 1967.

Sckopke, N., G. Paschmann, A. L. Brinca, C. W. Carlson and H. Luhr, Ion thermalization in quasiperpendicular shocks involving reflected ions, *J. Geophys. Res.*, **95**, 6337, 1990.

Shawhan, S. D., Magnetospheric plasma waves, in *Solar System Plasma Physics, Vol. III*, L. J. Lanzerotti, C. F. Kennel and E. H. Parker, Eds., North-Holland, 1979.

Song, P., R. C. Elphic, C. T. Russell, J. T. Gosling and Ca. A. Cattell, Structure and properties of the subsolar magnetopause for northward imf: ISEE observations, *J. Geophys. Res.*, **95**, 6375, 1990.

Tanaka, M., Simulations of heavy ion heating by electromagnetic ion cyclotron waves driven by proton temperature anisotropies, *J. Geophys. Res.*, **90**, 6459, 1985.

Thomsen, M. F., J. T. Gosling, S. J. Bame and M. M. Mellott, ion and electron heating at collisionless shocks near the critical Mach number, *J. Geophys. Res.*, **90**, 137, 1985.

Tsurutani, B. T., E. J. Smith, R. R. Anderson, K. W. Ogilvie, J. D. Scudder, D. N. Baker and S. J. Bame, Lion roars and nonoscillatory drift mirror waves in the magnetosheath, *J. Geophys. Res.*, **87**, 6060, 1982.

Winske, D., and K. B. Quest, Magnetic field and density fluctuations at perpendicular supercritical collisionless shocks, *J. Geophys. Res.*, **93**, 9681, 1988.

8

Electromagnetic component/component instabilities in uniform plasmas

In this chapter we continue to study electromagnetic fluctuations in homogeneous, magnetized, collisionless plasmas. The new element here is that we consider the zeroth-order distribution function of each plasma component to be Maxwellian with drift velocity $\mathbf{v}_{oj}$ parallel or antiparallel to $\mathbf{B}_o$ (Equation (3.1.3)). If two components have a relative drift $\mathbf{v}_o$ greater than some threshold, the corresponding free energy can lead to instability growth. Section 8.1 outlines the derivation of the dispersion equation for this case; Section 8.2 discusses electromagnetic ion/ion instabilities; Section 8.3 addresses electromagnetic electron/electron instabilities; Section 8.4 considers electromagnetic electron/ion instabilities; and Section 8.5 examines the consequences of electromagnetic effects on ion/ion instabilities that are electrostatic in the limit of zero β. Section 8.6 is a brief summary.

Space plasma heating and acceleration processes typically act on both species and are likely to give rise to beam/core distributions for both electrons and ions. However, in contrast to the case of $T_{\perp j} > T_{\parallel j}$ discussed in the previous chapter, the instabilities driven by beam/core free energies do not clearly separate into low frequency ion-driven and high frequency electron-driven modes. Thus, although we treat relative ion drifts and relative electron drifts separately in this chapter, this separation is due more to our desire to clarify the presentation than to any compelling physical arguments. Thus, in Sections 8.2 through 8.5, we consider a two-species, three-component plasma consisting of a relatively tenuous beam (denoted by subscript b), a relatively dense core (c), and a third component of the other species. The beam and core have drift velocities $\mathbf{v}_{ob}$ and $\mathbf{v}_{oc}$ that satisfy $\mathbf{v}_{oj} \times \mathbf{B}_o = 0$; we denote the relative beam/core component drift velocity by $\mathbf{v}_o = \mathbf{v}_{ob} - \mathbf{v}_{oc}$. We consider a charge neutral plasma $n_e = n_i = n_c + n_b$ and zero net current: $\sum_j e_j n_j \mathbf{v}_{oj} = 0$.

8.1 The electromagnetic dispersion equation

8.1.1 $\mathbf{k} \times \mathbf{B}_o \neq 0$

The derivation of the dispersion equation at oblique propagation again follows the procedure in Section 5.1 through Equation (5.1.8). For a drifting Maxwellian zeroth-order distribution (Equation (3.1.3)), the first order distribution function is, from Equation (5.1.8),

$$f_j^{(1)}(\mathbf{k},\mathbf{v},\omega) = \frac{e_j}{T_j} f_j^{(M)}(|\mathbf{v}-\mathbf{v}_{oj}|)\left(1-\frac{\mathbf{k}\cdot\mathbf{v}_{oj}}{\omega}\right)\int_{-\infty}^{0} d\tau \mathbf{v}'\cdot\mathbf{E}^{(1)}(\mathbf{k},\omega)\ \exp\left[ib_j(\tau,\omega)\right]$$

$$-\frac{e_j}{T_j} f_j^{(M)}(|\mathbf{v}-\mathbf{v}_{oj}|)\mathbf{v}_{oj}\cdot\mathbf{E}^{(1)}(\mathbf{k},\omega)\int_{-\infty}^{0} d\tau\left(1-\frac{\mathbf{k}\cdot\mathbf{v}'}{\omega}\right)\exp\left[ib_j(\tau,\omega)\right] \quad (8.1.1)$$

The dimensionless conductivity tensor is then, from Equation (5.1.1),

$$\mathbf{S}_j(\mathbf{k},\omega) =$$

$$\frac{ik_j^2}{k^2 n_j c^2}(\omega-\mathbf{k}\cdot\mathbf{v}_{oj})\int d^3v(\mathbf{v}+\mathbf{v}_{oj})f_j^{(M)}(v)\int_{-\infty}^{0} d\tau(\mathbf{v}'+\mathbf{v}_{oj})\exp[ib_j(\tau,\omega-\mathbf{k}\cdot\mathbf{v}_{oj})]$$

$$+\frac{k_j^2}{k^2}\frac{\mathbf{v}_{oj}\mathbf{v}_{oj}}{c^2}. \quad (8.1.2)$$

Thus

$$\mathbf{S}_j(\mathbf{k},\omega) = \mathbf{S}_j^{(M)}(\mathbf{k},\omega-\mathbf{k}\cdot\mathbf{v}_{oj}) + \frac{\mathbf{v}_{oj}}{c}\mathbf{Q}_j^{(M)}(\mathbf{k},\omega-\mathbf{k}\cdot\mathbf{v}_{oj})$$

$$+\mathbf{R}_j^{(M)}(\mathbf{k},\omega-\mathbf{k}\cdot\mathbf{v}_{oj})\frac{\mathbf{v}_{oj}}{c} + \frac{\mathbf{v}_{oj}\mathbf{v}_{oj}}{c^2}K_j^{(M)}(\mathbf{k},\omega-\mathbf{k}\cdot\mathbf{v}_{oj}) \quad (8.1.3)$$

where $\mathbf{S}_j^{(M)}(\mathbf{k},\omega)$ is the Maxwellian-based conductivity tensor of Equation (6.1.3),

$$\mathbf{Q}_j^{(M)}(\mathbf{k},\omega) = \frac{ik_j^2\omega}{k^2 c n_j}\int d^3v f_j^{(M)}(v)\int_{-\infty}^{0} d\tau \mathbf{v}' \exp[ib_j(\tau,\omega)] \quad (8.1.4)$$

$$\mathbf{R}_j^{(M)}(\mathbf{k},\omega) = \frac{ik_j^2\omega}{k^2 c n_j}\int d^3v \mathbf{v} f_j^{(M)}(v)\int_{-\infty}^{0} d\tau \exp[ib_j(\tau,\omega)] \quad (8.1.5)$$

and $K_j^{(M)}(\mathbf{k},\omega)$ is the electrostatic susceptibility of Equation (2.3.9).

Evaluating these integrals through the techniques of Appendix C, one obtains

$$\mathbf{Q}_j^{(M)}(\mathbf{k},\omega) = \frac{k_j^2}{k^2}\frac{v_j}{c}\frac{\omega}{\sqrt{2}|k_z|v_j}\left\{i\hat{\mathbf{x}}\frac{k_y v_j}{\Omega_j}\exp(-\lambda_j)\sum_{m=-\infty}^{\infty}[I_m(\lambda_j)-I_m'(\lambda_j)]Z(\zeta_j^m)\right.$$

$$-\hat{\mathbf{y}}\frac{\Omega_j}{k_y v_j}\exp(-\lambda_j)\sum_{m=-\infty}^{\infty} m I_m(\lambda_j) Z(\zeta_j^m) - \hat{\mathbf{z}}\frac{|k_z|}{\sqrt{2}k_z}\exp(-\lambda_j)\sum_{m=-\infty}^{\infty} I_m(\lambda_j) Z'(\zeta_j^m)\Bigg\}$$

$$R_{xj}^{(M)}(\mathbf{k},\omega) = -Q_{xj}^{(M)}(\mathbf{k},\omega)$$

$$R_{yj}^{(M)}(\mathbf{k},\omega) = Q_{yj}^{(M)}(\mathbf{k},\omega)$$

$$R_{zj}^{(M)}(\mathbf{k},\omega) = Q_{zj}^{(M)}(\mathbf{k},\omega)$$

and

$$K_j^{(M)}(\mathbf{k},\omega) = \frac{k_j^2}{k^2}\left[1 + \frac{\omega}{\sqrt{2}|k_z|v_j}\exp(-\lambda_j)\sum_{m=-\infty}^{\infty} I_m(\lambda_j) Z(\zeta_j^m)\right]$$

where $\lambda_j \equiv (k_y v_j/\Omega_j)^2$, $Z(\zeta_j^m)$ is the plasma dispersion function (Appendix A), and

$$\zeta_j^m \equiv \frac{\omega - \mathbf{k}\cdot\mathbf{v}_{oj} + m\Omega_j}{\sqrt{2}|k_z|v_j}.$$

8.1.2 $\mathbf{k}\times\mathbf{B}_o = 0$

In this case $k_y = 0$. Writing $k_z^2 = k^2$, the above expressions reduce to

$$\mathbf{Q}_j^{(M)}(\mathbf{k},\omega) = \mathbf{R}_j^{(M)}(\mathbf{k},\omega) = -\hat{\mathbf{z}}\frac{k_j^2}{k^2}\frac{v_j}{c}\frac{\omega}{2k_z v_j}Z'(\zeta_j^0)$$

and

$$K_j^{(M)}(\mathbf{k},\omega) = -\frac{k_j^2}{2k^2}Z'(\zeta_j^0)$$

Thus, as before, at $\mathbf{k}\times\mathbf{B}_o = 0$ Equation (5.1.2) factors into two parts. For electrostatic modes with $E_z^{(1)} \neq 0$, $D_{zz} = 0$ which, at $\omega^2 \neq 0$, yields the electrostatic dispersion equation for drifting Maxwellian distributions in an unmagnetized plasma, Equation (3.2.1) with Equation (3.2.3). Thus, for electrostatic modes propagating parallel to $\mathbf{B}_o$, we recover all of the results of Section 3.2.

From the remainder of the dispersion equation one obtains the electromagnetic dispersion equation for $\mathbf{k}\times\mathbf{B}_o = 0$:

$$\omega^2 - k^2c^2 + k^2c^2\sum_j S_j^{\pm}(\mathbf{k},\omega) = 0$$

where

$$S_j^{\pm}(\mathbf{k},\omega) = \frac{\omega_j^2}{k^2c^2}\zeta_j^0 Z(\zeta_j^{\pm 1}) \tag{8.1.6}$$

Note that this is the same form as Equation (6.1.7). The difference is in the definition of $\zeta_j^{\pm 1}$ which here contains the drift term $\mathbf{k} \cdot \mathbf{v}_{oj}$. In this case, the asymptotic limit of the plasma dispersion function yields the cold plasma dispersion equation

$$\omega^2 - k^2c^2 - \sum_j \frac{\omega_j^2(\omega - k_z v_{oj})}{\omega - k_z v_{oj} \pm \Omega_j} = 0 \qquad (8.1.7)$$

If $|\gamma| \ll |\omega_r|$ and one uses the drifting Maxwellian distribution of Equation (3.1.3) in Equation (5.1.16),

$$\gamma \simeq \frac{\pi}{2\omega_r} \sum_j \frac{\omega_j^2}{(2\pi v_j^2)^{1/2}} \left(v_{oj} - \frac{\omega_r}{k_z} \right) \exp\left[-\frac{(\omega_r \pm \Omega_j - k_z v_{oj})^2}{2k_z^2 v_j^2} \right]. \qquad (8.1.8)$$

The $(v_{oj} - \omega_r/k_z)$ factor is the same as one obtains from the analogous expression for electrostatic instabilities driven by drifting Maxwellians (e.g. Equation (3.2.4)), and implies that, in both cases, instabilities associated with this type of distribution have a threshold; i.e., v_{oj} of the driving species must be larger than ω_r/k_z. However, the condition necessary for an electrostatic instability, that there be a positive slope on an appropriate reduced distribution function at the phase speed of the wave, does not apply to electromagnetic fluctuations, as Equation (5.1.16) shows.

8.2 Electromagnetic ion/ion instabilities

In this section we consider a two-species, three-component plasma consisting of Maxwellian electrons, and two ion components represented by drifting Maxwellian distributions, a more dense core and a less dense beam. We concentrate on electromagnetic ion/ion instabilities associated with a relatively weak ($n_b \ll n_c$) and energetic ($v_{ob} \gg v_c$) beam. At $\mathbf{k} \times \mathbf{B}_o = 0$, there are three electromagnetic instabilities that may arise from this configuration (Gary *et al.*, 1984): The right-hand resonant, the nonresonant and left-hand resonant ion/ion instabilities. The ion cyclotron anisotropy instability of Section 7.2.2 can be driven by $T_{\perp b} > T_{\parallel b}$, but requires a bi-Maxwellian zeroth-order beam distribution with a relatively large anisotropy to be dominant. Thus we here emphasize the three electromagnetic ion/ion instabilities, with cyclotron resonant speeds as schematically indicated in Figure 8.1.

In discussing ion/ion instabilities it is convenient to define three distinct beam regimes: if the beam satisfies $0 < v_b \ll v_{ob}$ we call it a "cool" beam; if $v_{ob} \sim v_b$ the beam is "warm," and we term the beam "hot" if it satisfies $v_{ob} \ll v_b$.

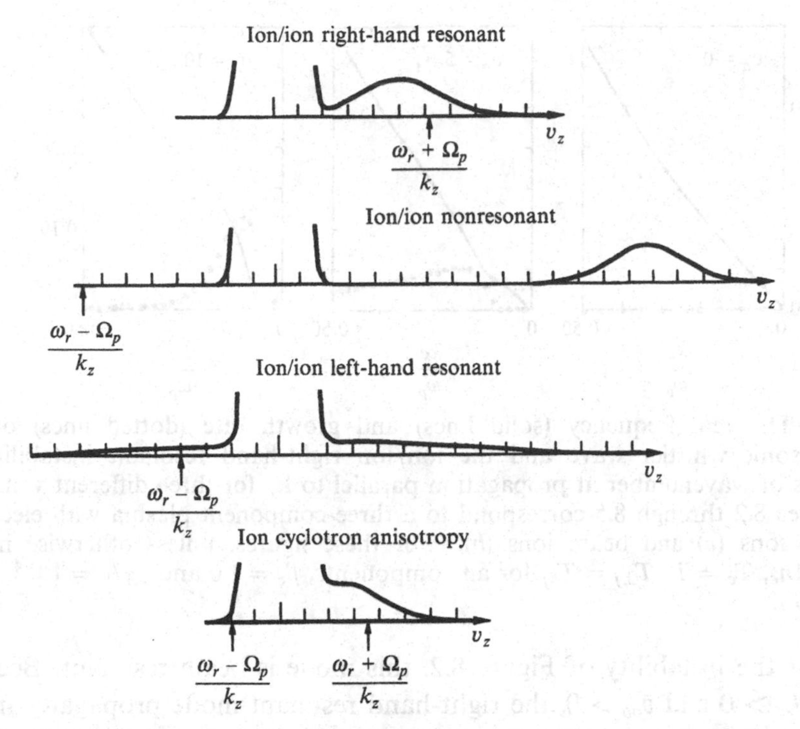

Fig. 8.1 Reduced zeroth-order ion distribution functions corresponding to four electromagnetic instabilities at $\mathbf{k} \times \mathbf{B}_o = 0$, and the associated cyclotron resonant speeds (From Fig. 1 of Gary and Tokar, 1985).

8.2.1 The ion/ion right-hand resonant instability

For a relatively isotropic, cool ion beam, the ion/ion right-hand resonant instability is the growing mode of lowest threshold (Barnes, 1970). Its dispersion properties are depicted in Figure 8.2. At $v_o = 0$ the mode of Figure 8.2 is the right-hand circularly polarized, positive helicity magnetosonic/whistler wave: it satisfies $\omega_r = kv_A$ in the long wavelength limit and makes the transition to whistler dispersion as the wavenumber increases. The right-hand instability that evolves from this wave becomes strongly dispersive at long wavelengths and, near wavenumbers of maximum growth, satisfies

$$\omega_r \simeq k_z v_{ob} - \Omega_i \tag{8.2.1}$$

At $\mathbf{k} \times \mathbf{B}_o = 0$, the $m = \pm 1$ cyclotron resonance is the only one possible for electromagnetic instabilities so that wave-particle interactions are strong for the jth species when $|\zeta_j^{\pm 1}| \lesssim 1$. For this instability, both the electrons and the core ions are nonresonant ($|\zeta_e^{+1}| \gg |\zeta_c^{+1}| \gg 1$). But because $|\zeta_b^{+1}|$ is of order

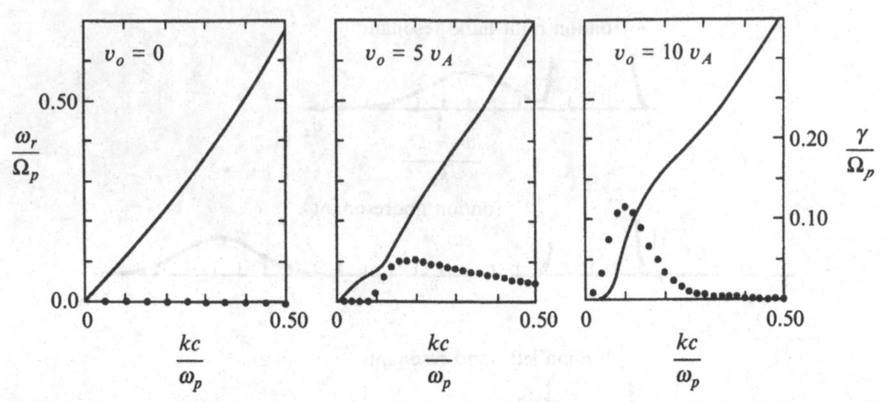

Fig. 8.2 The real frequency (solid lines) and growth rate (dotted lines) of the magnetosonic/whistler wave and the ion/ion right-hand resonant instability as functions of wavenumber at propagation parallel to $\mathbf{B}_o$ for three different values of v_o. Figures 8.2 through 8.5 correspond to a three-component plasma with electrons (e), core ions (c) and beam ions (b). For these figures, unless otherwise noted, $n_b = 0.01 n_e$, $T_e = T_c$, $T_{\perp j} = T_{\parallel j}$ for all components, $\beta_c = 1.0$ and $v_A/c = 10^{-4}$. Here $T_b = 10 T_c$.

unity for the instability of Figure 8.2, this mode is beam resonant. Because $\omega_r > 0, k_z > 0$ and $v_{ob} > 0$, the right-hand resonant mode propagates in the direction of the beam.

Numerical solutions of the linear dispersion equation at $\mathbf{k} \times \mathbf{B}_o = 0$ for $n_b \ll n_e$ and $v_A \ll v_{ob}$ show that near maximum growth the right-hand resonant instability approximately satisfies Equation (8.2.1). In addition, for beam densities in the range $0.01 \lesssim n_b/n_e \lesssim 0.10$, there is $\omega_r \simeq \gamma$ near the wavenumber of maximum growth. Then it follows that, for the ion/ion right-hand resonant instability,

$$\frac{\gamma_m}{\Omega_i} \simeq \left(\frac{n_b}{2n_e} \right)^{1/3} \tag{8.2.2}$$

Problem 8.2.1. Show that, under the conditions stated in the above paragraph, the imaginary part of the cold plasma dispersion equation (8.1.7) yields Equation (8.2.2).

8.2.2 *The ion/ion left-hand resonant instability*

For a sufficiently hot beam, there emerges another electromagnetic ion/ion instability illustrated in Figure 8.3 (Sentman *et al.*, 1981). This figure traces the evolution of the negative helicity, positive frequency mode as the beam velocity is increased from zero to $2v_A$ at $T_b = 100 T_c$. As v_o is increased, this

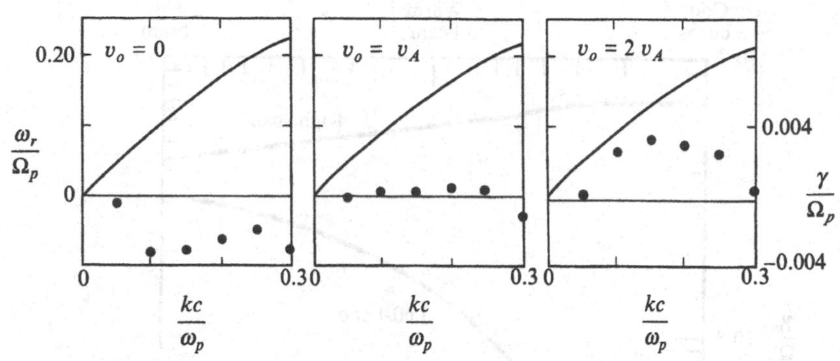

Fig. 8.3 The real frequency (solid lines) and damping/growth rate (dotted lines) of the Alfvén/ion cyclotron wave and the ion/ion left-hand resonant instability as functions of wavenumber at propagation parallel to $\mathbf{B}_o$ for three different values of v_o. Here $T_b = 100T_c$.

instability evolves out of the $\omega_r > 0$ Alfvén/ion cyclotron wave; because the real frequency remains positive for the modest growth rates associated with the instability, this mode is left-hand polarized and propagates parallel to the beam. Because $|\zeta_b^{-1}| < 1$ near maximum growth rate, Gary *et al.* (1984) call this the ion/ion left-hand resonant instability. The electrons and core ions are, as for the right-hand instability, nonresonant ($|\zeta_e^{+1}| \gg |\zeta_c^{+1}| \gg 1$).

If the ion beam is isotropic in its own frame, the thresholds of both the right-hand resonant and left-hand resonant instabilities lie near $v_o = v_A$ (e.g. Gary, 1985a). However, as is evident from Figure 8.1, under cool beam conditions, there are essentially no ions to resonate with the left-hand mode, and the right-hand mode has the larger growth rate. As the beam temperature increases, or as the plasma β increases, more backward-travelling ions come into resonance, and the two modes can have similar growth rates. Figure 8.4 illustrates this point by showing the maximum growth rate of the two beam resonant modes as a function of β_c.

Problem 8.2.2. Explain why, although T_b/T_c is constant in Figure 8.4, the increasing β_c brings more backward-travelling ions into resonance with the left-hand polarized mode and permits the maximum growth rate of this instability to increase.

Several studies of electromagnetic ion/ion resonant instabilities at oblique propagation have been carried out (Gary *et al.*, 1984; Smith *et al.*, 1985; Brinca and Tsurutani, 1989a), and, in almost every case, the fastest growing instability appears at $\mathbf{k} \times \mathbf{B}_o = 0$. Oblique propagation, however, does yield the interesting phenomenon of local (in **k**-space) maxima in the growth rate

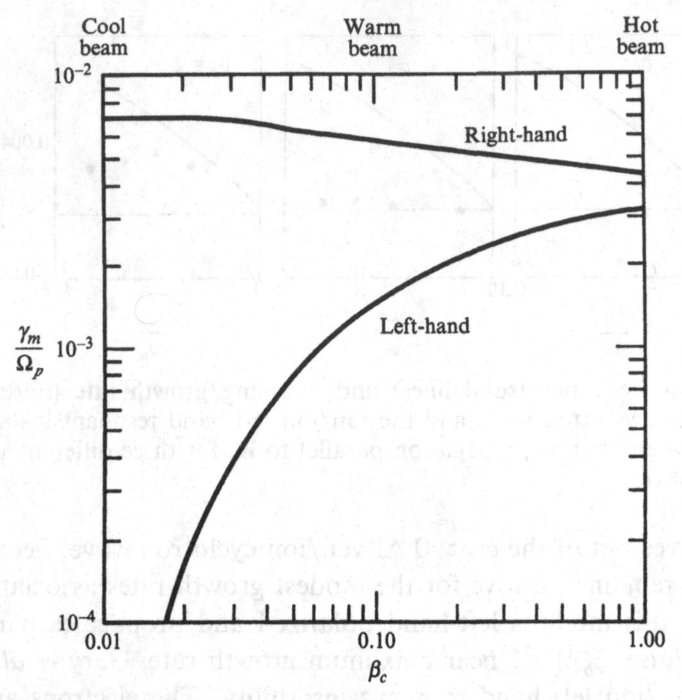

Fig. 8.4 The maximum growth rates of the ion/ion right-hand resonant and left-hand resonant instabilities as functions of the ion core β. Here $T_b = 100T_c$ and $v_o = 2v_A$ (From Fig. 7 of Gary, 1991).

at $n = 2, 3, ...$ harmonics of the proton cyclotron resonance (Goldstein *et al.*, 1983, but also see Goldstein *et al.*, 1985; Brinca and Tsurutani, 1989b).

8.2.3 The ion/ion nonresonant instability

A cool fast ion beam can also excite a firehose-like instability for which all three components are nonresonant ($|\zeta_j^{-1}| \gg 1$ for $j = c, b, e$; see Figure 8.1). Unlike the two ion/ion resonant instabilities, this nonresonant mode propagates in the direction opposite to the beam (Sentman *et al.*, 1981). Figure 8.5 illustrates the dispersion properties of this instability, which has negative helicity and a relatively small phase speed. The latter property implies that a relatively small change in the frame of reference can yield a change in the sense of polarization, so that the polarization is not a useful identifier for this instability.

Gary *et al.* (1984) demonstrated that the right-hand resonant instability typically has the lower threshold v_o, but that the nonresonant mode can have the larger growth rate if v_o/v_A and n_b/n_e are sufficiently large. This can

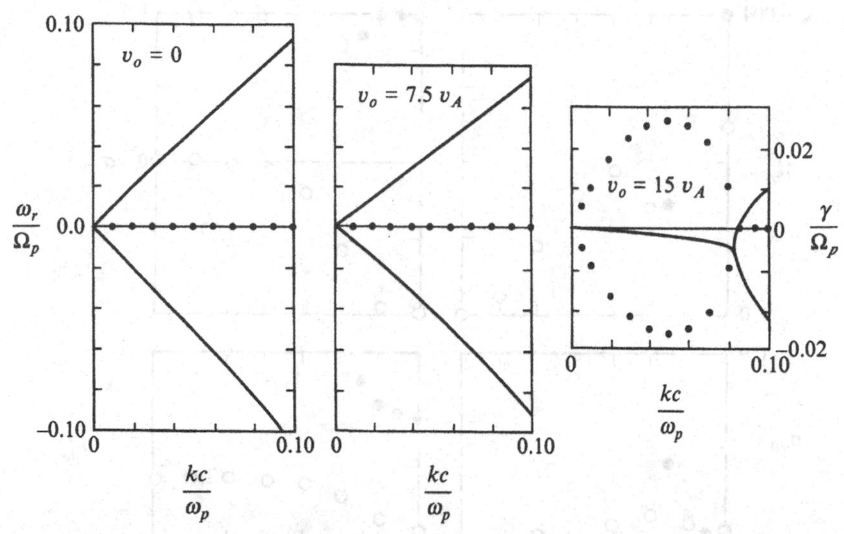

Fig. 8.5 The real frequency (solid lines) and damping/growth rates (dotted lines) of negative helicity waves and the ion/ion nonresonant instability as functions of wavenumber at $\mathbf{k} \times \mathbf{B}_o = 0$ for three different values of the ion beam/core relative drift speed. Here $T_b = 10T_c$.

be made clear by comparing Equation (8.2.2) with the Winske and Leroy (1984) expression for the maximum growth rate of the ion/ion nonresonant instability at $n_b \ll n_e$ and $v_A \ll v_o$:

$$\frac{\gamma_m}{\Omega_i} \simeq \frac{n_b}{2n_e}\frac{v_o}{v_A} \tag{8.2.3}$$

Winske and Gary (1986) also have demonstrated that the ion/ion nonresonant instability is more likely to have a larger growth rate if the beam is composed of ions heavier than protons.

8.2.4 Transport ratios for ion/ion instabilities

Figure 8.6 compares the growth rates and five transport ratios of the right-hand resonant and left-hand resonant instabilities as a function of propagation angle at $kc/\omega_p = 0.10$ and $T_b = 100T_h$. Both modes have maximum growth rate at parallel propagation for these parameters.

Just as $\omega(\mathbf{k})$ of the left-hand instability does not show much change from the zero-drift Alfvén/ion cyclotron wave (Figure 8.3), so the five transport ratios do not show much change from the corresponding values of the single proton component case (compare Figure 6.2). In contrast to fluctuations associated with isotropic distributions, however, the helicity and cross-helicity

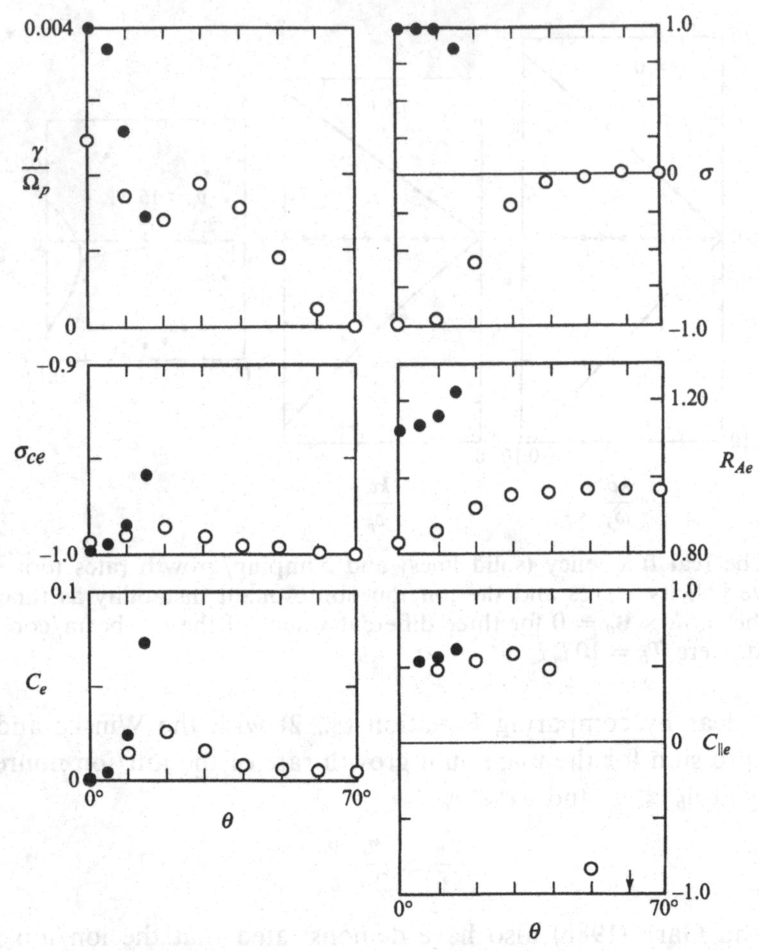

Fig. 8.6 The growth rate and five transport ratios of the ion/ion right-hand resonant instability (solid dots) and the ion/ion left-hand resonant instability (open dots) as functions of propagation angle at $kc/\omega_p = 0.10$. Here $T_b = 100T_c$, $v_{ob} > 0$, and $v_o = 2v_A$.

have unique signs associated with the direction of beam propagation. If the beam propagates parallel to $\mathbf{B}_o$, then $\omega_r > 0$; then, as shown in figure 8.6, it follows that $\sigma_{ce} < 0$ for both modes, but $\sigma > 0$ for the right-hand instability with $\sigma < 0$ for the left-hand instability. If the beam propagates antiparallel to the background magnetic field, then $\omega_r < 0$ and the signs of both σ and σ_{ce} are reversed for both instabilities. Thus, if one can determine the sign of $\mathbf{v}_{ob} \cdot \mathbf{B}_o$, then the fluctuation helicity may be used to discern the difference between the right-hand and left-hand resonant instabilities.

Figure 8.7 compares the growth rates and five transport ratios of the right-hand resonant and nonresonant instabilities as a function of propagation

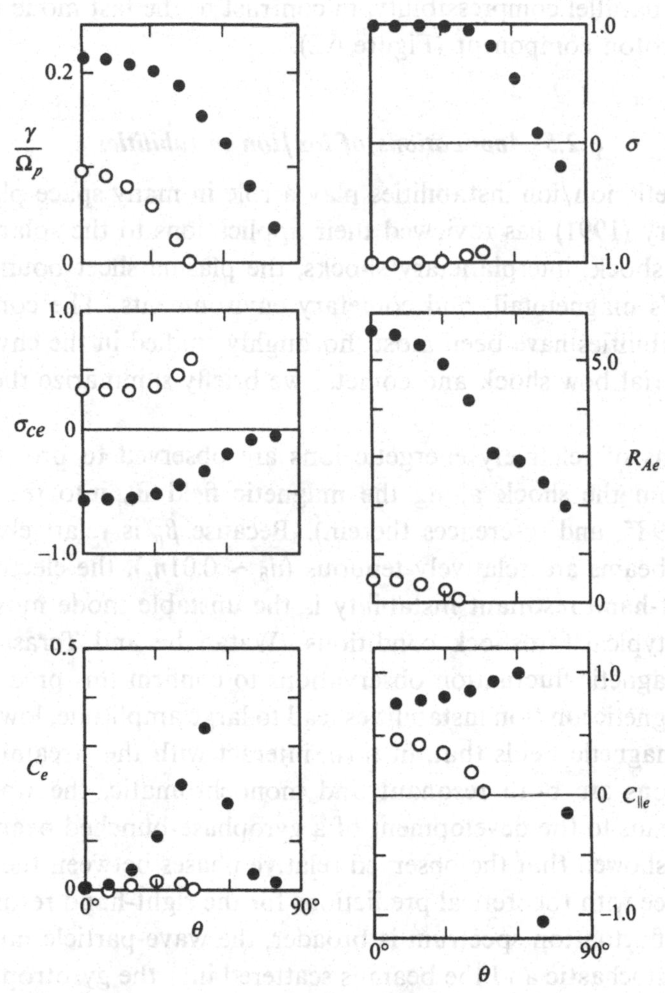

Fig. 8.7 The growth rate and five transport ratios of the ion/ion right-hand resonant instability (solid dots) and the ion/ion nonresonant instability (open dots) as functions of propagation angle at $kc/\omega_p = 0.15$. Here $T_b = 10T_c$, $v_{ob} > 0$, and $v_o = 8v_A$.

angle at $kc/\omega_p = 0.15$ and $T_b = 10T_h$. Both modes have maximum growth rate at parallel propagation for these parameters. Because this fast, cool beam regime corresponds to significant departures from the magnetosonic and Alfvén $\omega(\mathbf{k})$, we may expect that the transport ratios should also show changes from the single proton component case, and that is indeed true here. In contrast to the Alfvén/ion cyclotron wave and the left-hand resonant instability of the preceeding figure, the ion/ion nonresonant instability bears a positive cross-helicity and a relatively small Alfvén ratio. And the right-hand resonant instability in the present cool, fast beam regime has, at large

θ, a negative parallel compressibility, in contrast to the fast mode in the case of a single proton component (Figure 6.2).

8.2.5 Applications of ion/ion instabilities

Electromagnetic ion/ion instabilities play a role in many space plasma phenomena; Gary (1991) has reviewed their applications to the solar wind, the Earth's bow shock, interplanetary shocks, the plasma sheet boundary layer of the Earth's magnetotail, and cometary environments. The contributions of these instabilities have been most thoroughly studied in the environments of the terrestrial bow shock and comets; we briefly summarize these studies here.

Cool beams of relatively energetic ions are observed to propagate back upstream from the shock along the magnetic field $\mathbf{B}_o$ into the foreshock (Thomsen, 1985, and references therein). Because β_p is relatively high and because the beams are relatively tenuous ($n_b \sim 0.01 n_e$), the electromagnetic ion/ion right-hand resonant instability is the unstable mode most likely to arise under typical foreshock conditions; Watanabe and Terasawa (1984) have used magnetic fluctuation observations to confirm this prediction.

Electromagnetic ion/ion instabilities lead to large amplitude, low frequency fluctuating magnetic fields that, in turn, interact with the streaming ions. If the fluctuations are both resonant and monochromatic, the wave-particle interaction leads to the development of a gyrophase-bunched beam; Fuselier *et al.* (1986) showed that the observed relative phases between the beam and the wave agree with theoretical predictions for the right-hand resonant instability. If the fluctuation spectrum is broader, the wave-particle interaction is more nearly stochastic and the beam is scattered into the gyrotropic intermediate distributions; in either case, the end product is the strongly scattered, hot, diffuse component (Winske and Leroy, 1984). Competing models for the source of diffuse distributions in the terrestrial foreshock have been reviewed by Gosling *et al.* (1989); that paper shows two diffuse events that are very likely due to scattering by the co-existing magnetic fluctuations.

"Hot diamagnetic cavities," localized regions of the foreshock in which solar wind flow kinetic energy appears to have been converted to plasma thermal energy (Thomsen *et al.*, 1988, and references therein), may also be related to electromagnetic ion/ion instabilities. Thomsen *et al.* (1988) suggested that these features may form from strong interactions between shock-reflected ions and the solar wind mediated by the ion/ion nonresonant instability. Two-dimensional hybrid simulations by Thomas and Brecht (1988) have supported this hypothesis; in these computations an ion beam of finite extent yields growing fluctuations; the fluctuations heat the plasma that, in turn, pushes aside the ambient magnetic field. One problem is that Thomas and Brecht appear to see the right-hand resonant, rather than the nonresonant, instability. The one-dimensional hybrid simulations of Galvez *et al.* (1990) have provided further support for the Thomsen *et al.* (1988) hypothesis by showing that only the nonresonant instability can provide the strong alpha particle heating that is observed in these regions.

Quasiparallel shocks, and the possible role of electromagnetic ion/ion instabilities, have recently been the subject of several studies reviewed by Krauss-Varban and Omidi (1991).

A cometary nucleus sublimates neutral gas that freely expands into nearby space. When a neutral atom or molecule becomes ionized, for example by solar radiation, the resulting ion and electron immediately begin to interact with the magnetic field of the solar wind and the newborn cometary ions in particular become sources of free energy for a variety of instabilities. The nature of this free energy is a function of α, the angle between the solar wind velocity $\mathbf{v}_{sw}$ and the interplanetary magnetic field $\mathbf{B}_o$. At $\alpha = 0°$, cometary ions form a cold beam; at $\alpha = 90°$, they constitute a ring distribution, and at intermediate values of α they are described by a ring-beam distribution (Wu and Davidson, 1972).

Spacecraft observations of cometary environments show the presence of low frequency ($\omega_r \ll \Omega_p$) magnetic fluctuations that are moderately enhanced over the solar wind background near comet Halley (Glassmeier *et al.*, 1989, and references therein) and strongly enhanced at comet Giacobini-Zinner (Tsurutani and Smith, 1986). These observations have stimulated an extensive literature on the linear theory of electromagnetic instabilities that may be excited in distant cometary environments as α ranges from 0° to 90°; reviews of this research include Brinca and Tsurutani (1988) and Gary (1991). Linear theory has predicted, and computer simulations (Gary *et al.*, 1989) have verified, that, for a tenuous cometary ion density, the ion/ion right-hand resonant instability is the dominant growing mode at $0° \leq \alpha \lesssim 45°$. At larger values of α, the ring-like nature of the ring-beam distribution begins to assert itself, and left-hand polarized ion cyclotron anisotropy instabilities become important. The exact α value at which the left-hand instabilities attain the larger growth rate appears to depend on the model distribution; a typical value is the $\alpha \simeq 75°$ quoted by Thorne and Tsurutani (1987).

Although linear instability growth rates generally increase as α increases (Brinca and Tsurutani, 1988; Gary and Madland, 1988), computer simulations typically yield a lower level of saturation for electromagnetic ion anisotropy instabilities than for electromagnetic ion/ion instabilities (see

Gary and Winske, 1990, for comparative references). Computer simulations that model the distant cometary environment through the continual injection of cometary ions (Gary *et al.*, 1989; Miller *et al.*, 1991) do indeed show a decrease in fluctuation level as α increases. This result correlates well with the comet Giacobini-Zinner observations of Tsurutani *et al.* (1989) who report the absence of large amplitude magnetic fluctuations near the water-group ion cyclotron frequency for conditions such that α is near 90°.

Recent computer simulations have also addressed the cometary ion pitch-angle scattering that results from the growth of these electromagnetic instabilities. The results show that, although the fluctuating field growth rate diminishes as α increases toward 90°, the pitch-angle scattering rate of water-group cometary ions actually increases as α approaches the perpendicular (Gary *et al.*, 1991). These results, which are in qualitative agreement with the increase in cometary ion pitch-angle scattering observed by Neugebauer *et al.* (1990), are not yet explained.

8.3 Electromagnetic electron/electron instabilities

In this section we consider a two-species, three-component plasma consisting of ions, a tenuous electron beam (denoted by subscript b) and a more dense core electron component (c). At $\mathbf{k} \times \mathbf{B}_o = 0$, there are two electromagnetic electron/electron instabilities that may arise from this configuration (Gary, 1985b): the right-hand polarized whistler heat flux instability (Gary *et al.*, 1975a, 1975b) and the electron/electron firehose instability. Gary (1985b) has demonstrated that, for a very broad parameter range, the former mode generally has the lower threshold; moreover, that same paper has argued that, at sufficiently large β_i and small beam density, the whistler heat flux mode has a much lower threshold than either the magnetosonic or Alfvén heat flux instabilities at propagation oblique to $\mathbf{B}_o$. Thus we concentrate on the whistler heat flux instability in this section.

8.3.1 The whistler heat flux instability

Gary (1975b) has shown that this mode has maximum growth rate at parallel propagation, so it is sufficient to use the $\mathbf{k} \times \mathbf{B}_o = 0$ dispersion equation. The dispersion properties of this mode are illustrated in Figure 8.8, which traces $\omega(k)$ as the beam velocity is increased from zero. At $v_o = 0$ the mode shown is the right-hand circularly polarized magnetosonic/whistler wave. However, the whistler heat flux instability that evolves from this mode exhibits a decrease in ω_r as the beam drift speed increases. So, although the mode

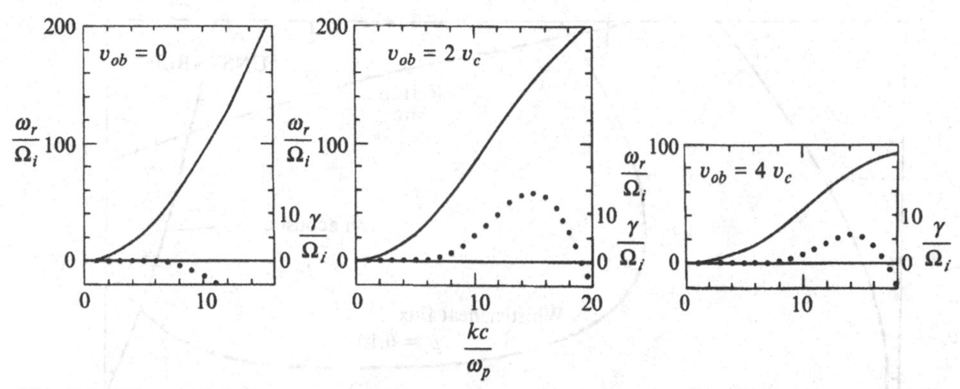

Fig. 8.8 The real frequency (solid lines) and damping/growth rate (dotted lines) of the whistler mode and the whistler heat flux instability as functions of wavenumber at propagation parallel to $\mathbf{B}_o$ for three different values of v_{ob}. Figures 8.8 through 8.10 correspond to a three-component plasma with ions (i), beam electrons (b) and core electrons (c); unless otherwise noted, $n_b = 0.10n_e$, $T_b = 10T_c$, $T_i = T_c$, $T_{\perp j} = T_{\parallel j}$ for all components, $\beta_c = 1.0$ and $\omega_e/|\Omega_e| = 100.0$.

remains whistler-like in the sense that $\Omega_i < \omega_r < |\Omega_e|$, the real frequency no longer well satisfies the cold plasma dispersion equation. For this instability, the ions are nonresonant ($|\zeta_i^{+1}| \gg 1$), the core component electrons are weakly resonant ($|\zeta_c^{+1}| > 1$) and the beam electrons are resonant ($|\zeta_b^{+1}| \sim 1$).

The threshold drift speeds of the whistler heat flux instability, the electron/ion acoustic instability and electron/electron beam instability are plotted as functions of the beam density in Figure 8.9. The electron/electron beam and ion acoustic instabilities are electrostatic and hence independent of β_i. The whistler heat flux instability is electromagnetic and a sensitive function of β_i, so its threshold is plotted for two different values of that parameter. This figure illustrates what is more completely documented in Gary (1985b): for sufficiently large $\beta_i, n_b/n_e$ and T_b/T_c, the whistler heat flux instability has the lowest threshold of any electron beam instability, electrostatic or electromagnetic.

Problem 8.3.1. Show that the electron/electron drift speed thresholds of Figure 8.9 correspond to $v_o \gg v_A$. Explain why the whistler instability typically has much higher drift speed thresholds than the electromagnetic ion/ion instabilities.

The maximum growth rate of the whistler heat flux instability is enhanced by increases in n_b/n_e, β_i and, as illustrated in Figure 8.10, T_b/T_c. The electrons most strongly in resonance with the wave satisfy $v_z = (\omega_r + \Omega_e)/k_z$; because $\omega_r \ll |\Omega_e|$ and $\Omega_e < 0$, it is those beam electrons moving opposite

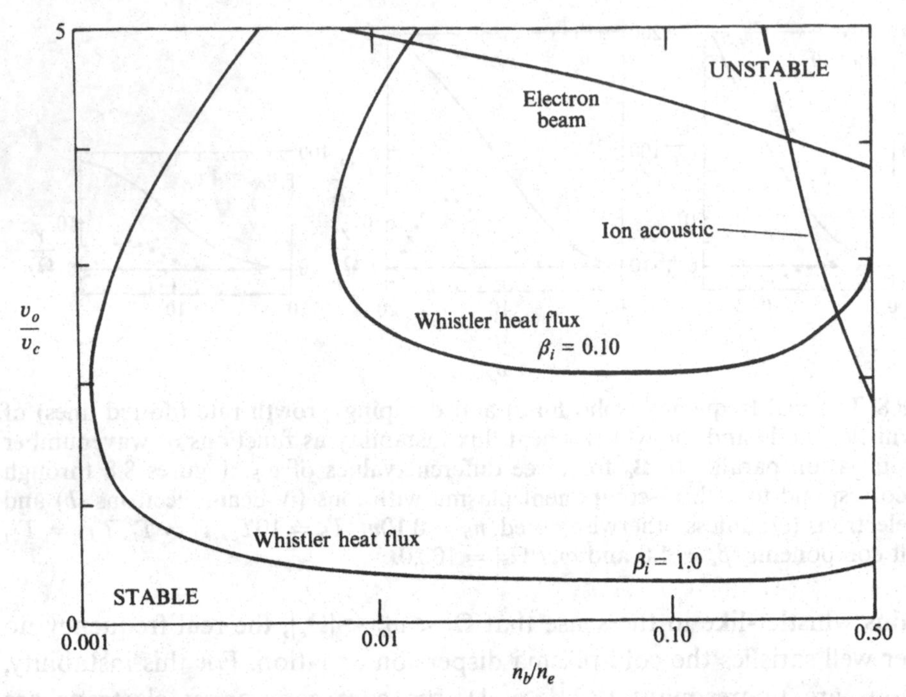

Fig. 8.9 The threshold beam/core drift speeds for three different instabilities as functions of the beam density. The threshold of the electron/electron beam and electron/ion acoustic instabilities correspond to $\gamma > 0$; the whistler heat flux instability curves correspond to $\gamma \geq 10^{-1}\Omega_i$. The areas above the various curves are unstable or more unstable to the associated instabilities; the areas below are stable or more nearly stable (From Fig. 7 of Gary, 1985b).

to v_{ob} and ω_r/k_z that drive the instability. Just as with the ion/ion left-hand resonant instability (see Figure 8.1), an increasing beam temperature puts more of these particles in resonance with the wave and enhances the instability. Thus the whistler heat flux instability requires a relatively warm or hot beam with $v_b \gtrsim v_{ob}$. Another feature of this "backward" resonance is that, as v_o becomes larger, the number of resonant particles is eventually depleted, so that γ_m reaches a maximum and then returns to zero at large drift speeds, as shown in Figure 8.10.

8.3.2 Applications of electron/electron instabilities

The results of Subsection 8.3.1 have shown that electromagnetic electron/electron instabilities require hot electrons beams and plasma β values of order unity in order to compete with their electrostatic counterparts. Because hot electron beams are not frequently observed in the magnetosphere, studies of

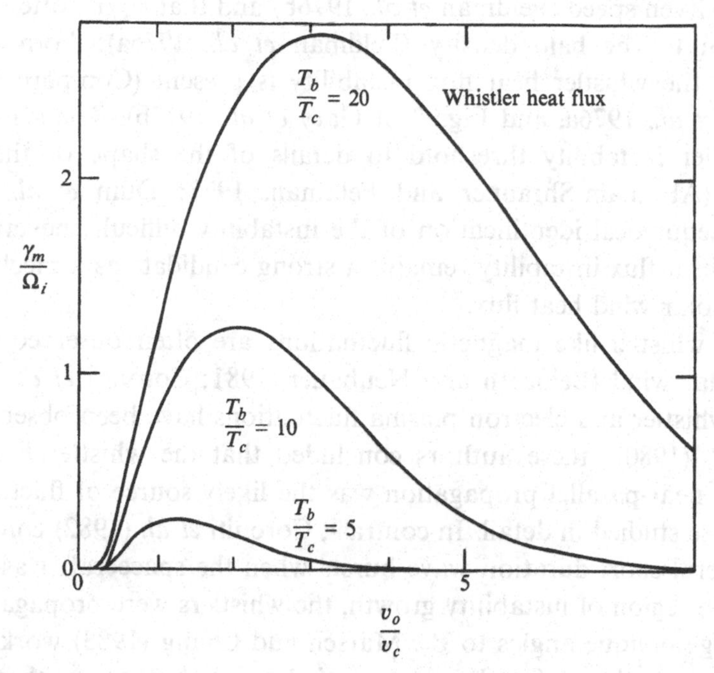

Fig. 8.10 The maximum growth rate as a function of beam/core relative drift for the whistler heat flux instability at three different values of T_b/T_c (From Fig. 5 of Gary, 1985b).

electron/electron whistler instabilities have been directed primarily toward the bow shock (Tokar *et al.*, 1984) and the solar wind (Schwartz, 1980, and references therein). Here we summarize studies of the whistler heat flux instability in the latter medium.

Detailed observations (Feldman *et al.*, 1975, and references therein) have demonstrated that solar wind electron distributions near 1 AU are well characterized by a superposition of two bi-Maxwellian distributions with relative streaming velocities parallel to $\mathbf{B}_o$ and satisfying the zero current condition. The velocity difference between the cooler, more dense core and the hotter, more tenuous halo corresponds to a heat flux in the solar wind frame and is a potential free energy source for several different electron/electron or electron/ion instabilities (Schwartz, 1980).

The $T_e \sim T_p$, $\beta_p \sim 1$ conditions of the solar wind indicate that the whistler heat flux instability often has the lowest threshold drift speed by comparison not only with other electromagnetic heat flux instabilities (Gary 1975a, 1975b), but also with respect to the electrostatic electron/ion acoustic instability (Gary, 1978). Further analysis of solar wind electron distributions has shown that the core/proton relative drift speed often varies in proportion to the local Alfvén speed (Feldman *et al.*, 1976b) and that v_{oc}/v_A often varies in proportion to the halo density (Feldman *et al.*, 1976a); both results indicate that the whistler heat flux instability is present (Compare Fig. 7 of Feldman *et al.*, 1976a, and Fig. 7 of Gary *et al.*, 1975b). The sensitivity of the whistler instability threshold to details of the shape of the halo distribution (Abraham-Shrauner and Feldman, 1977; Dum *et al.*, 1980) makes an unequivocal identification of the instability difficult; nevertheless the whistler heat flux instability remains a strong candidate as a mechanism to limit the solar wind heat flux.

Enhanced whistler-like magnetic fluctuations are often observed in the disturbed solar wind (Beinroth and Neubauer, 1981; Coroniti *et al.*, 1982). Correlated whistler and electron plasma fluctuations have been observed by Kennel *et al.* (1980); these authors concluded that the whistler heat flux instability at near-parallel propagation was the likely source of fluctuations in the one case studied in detail. In contrast, Coroniti *et al.* (1982) concluded that, for several short-duration wave bursts when the spacecraft was possibly within the region of instability growth, the whistlers were propagating at relatively large oblique angles to $\mathbf{B}_o$. Marsch and Chang (1983) worked out a linear dispersion theory for electromagnetic lower hybrid waves that could be driven by the electron core/halo relative drift and propagate nearly perpendicular to the interplanetary magnetic field, thereby providing a possible explanation for the observations of Coroniti *et al.*

8.4 Electromagnetic electron/ion instabilities

Although the linear dispersion properties of electromagnetic electron/ion instabilities associated with a nonzero current have been established (Gary *et al.*, 1976), computer simulations often show that these modes saturate at very low fluctuation amplitudes (Gary and Winske, 1990). Thus, such instabilities have not received much attention with respect to space plasma applications. However, electromagnetic electron/ion instabilities can arise under zero current conditions in either an electron/electron or an ion/ion streaming configuration, and there are potential space plasma applications in both such cases.

If a magnetic-field-aligned ion beam is present and the zero current condition is satisfied in a plasma, then the relative electron/ion-core and electron/ion-beam drifts constitute additional sources of free energy beyond the ion-core/ion-beam drift speed. Akimoto *et al.* (1987) studied the linear theory of electron/ion whistler instabilities in the ion/ion configuration, and found two related but distinct growing modes, one at $\mathbf{k} \times \mathbf{B}_o = 0$ and one

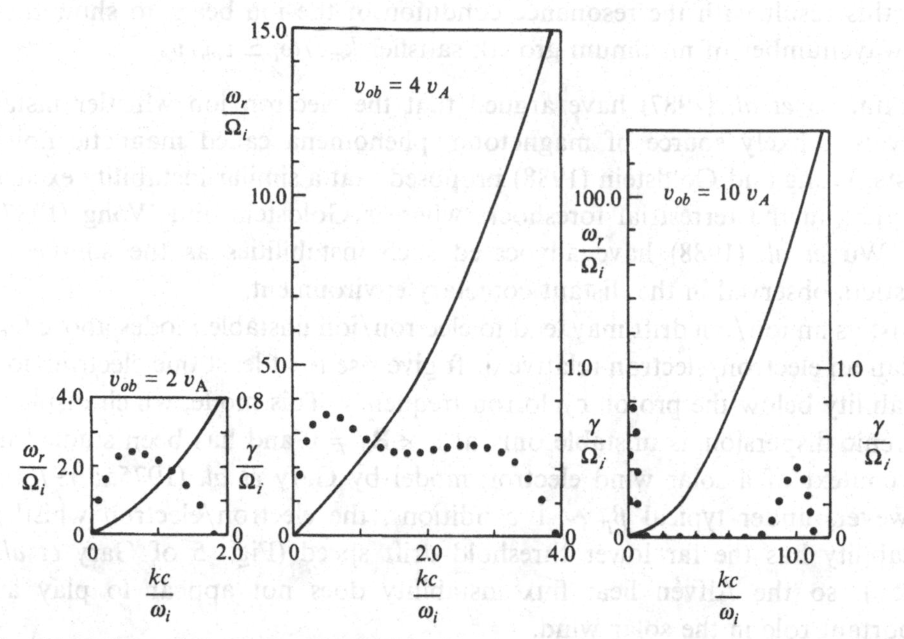

Fig. 8.11 The real frequency (solid lines) and growth rates (dotted lines) of positive helicity modes as functions of wavenumber at propagation parallel to $\mathbf{B}_o$ for three different values of the electron beam drift speed. This figure corresponds to a three-component plasma with ions (*i*), beam electrons (*b*) and core electrons (*c*); $n_b = 0.60n_e$, $T_b = T_c$, $T_i = T_c$, $T_{\perp j} = T_{\parallel j}$ for all components, $\beta_c = 1.0$ and $\omega_e/|\Omega_e| = 20.0$ (From Fig. 1 of Akimoto *et al.*, 1987).

at oblique propagation. If $v_o \gtrsim v_A$, an ion/ion relative drift leads to growth of the ion/ion right-hand resonant instability at $\omega_r < \Omega_i$ and $kc/\omega_i < 1$. As the beam/core relative drift is increased, Figure 8.11 shows that the range of unstable frequencies and wavenumbers increases so that the instability moves up the whistler dispersion curve. At $v_o \gg v_A$ and parallel propagation, the unstable regime of the ion/ion instability collapses to $kc/\omega_i \ll 1$, whereas the electron/ion parallel whistler instability develops its own detached domain of unstable wavenumbers in the opposite limit. At parallel propagation, the core ions are nonresonant and the electrons are weakly resonant ($|\zeta_c^{+1}| \gg 1$ and $|\zeta_e^{+1}| > 1$), but the ion beam remains resonant through the approximate satisfaction of the cyclotron resonance condition (8.2.1).

Problem 8.4.1. Use the $v_o \gg v_A$ properties of Figure 8.11 to argue that the ion terms do not contribute to the electron/ion parallel whistler instability so that

$$\omega_r \simeq \frac{k^2c^2\Omega_i}{\omega_i^2}$$

Use this result with the resonance condition of the ion beam to show that the wavenumber of maximum growth satisfies $k_m c/\omega_i \simeq v_{ob}/v_A$.

Akimoto *et al.* (1987) have argued that the electron/ion whistler instability is a likely source of magnetotail phenomena called magnetic noise bursts. Wong and Goldstein (1988) proposed that a similar instability excites whistlers in the terrestrial foreshock, whereas Goldstein and Wong (1987) and Wu *et al.* (1988) have advocated such instabilities as the source of whistlers observed in the distant cometary environment.

Just as an ion/ion drift may lead to electron/ion unstable modes above Ω_p, so can an electron/electron relative drift give rise to at least one electron/ion instability below the proton cyclotron frequency. This mode, which displays Alfvénic dispersion, is unstable only at $\mathbf{k} \times \mathbf{B}_o \neq 0$ and has been studied in the context of a solar wind electron model by Gary *et al.* (1975a; 1975b). However, under typical $\beta_p \sim 1$ conditions, the electron/electron whistler instability has the far lower threshold drift speed (Fig. 5 of Gary *et al.*, 1975b), so the Alfvén heat flux instability does not appear to play an important role in the solar wind.

8.5 Electrostatic ion/ion instabilities: electromagnetic effects

Because $\beta_p = 2v_p^2/v_A^2$ and $2v_p^2 \lesssim c_s^2$, the relative drift speed at thresholds of electromagnetic ion/ion instabilities is typically smaller than at thresholds of electrostatic ion/ion instabilities at high β; the converse is true at $\beta \ll 1$. Most studies of the former modes, as in Section 8.2, have assumed relatively large values of β and have ignored the latter instabilities; many examinations of electrostatic ion/ion cyclotron instabilities, as in Section 3.3.2, have been carried out at zero β and have ignored the corresponding electromagnetic effects. In this section we examine how nonzero β and the resulting electromagnetic effects alter the properties of the electrostatic ion/ion cyclotron instability. Note that the ion/ion acoustic instability at $\mathbf{k} \times \mathbf{B}_o = 0$ is unaffected by the magnetic field and hence by increases in β.

The model considered here is the same as that used in Section 8.2: Maxwellian electrons, a relatively tenuous proton beam and a more dense proton core; the two proton components have a relative ion/ion drift speed parallel to $\mathbf{B}_o$. In this model, the electrostatic ion/ion cyclotron instability has a limited parameter regime in which it can both grow and maintain its status as the mode of most rapid linear growth (e.g. see Figure 3.20). Thus, in order to consider the ion/ion cyclotron instability in a regime in which it does not compete with other unstable modes, we choose a modest T_e/T_c

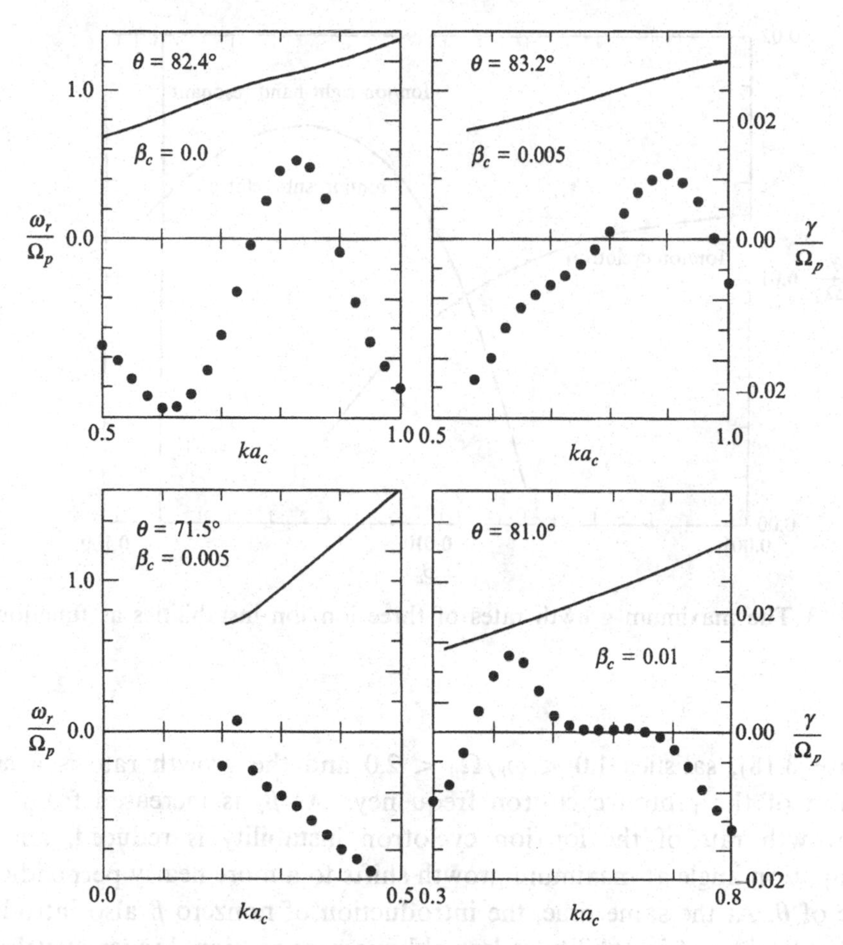

Fig. 8.12 The real frequency (solid lines) and damping/growth rate (dotted lines) of ion/ion instabilities as functions of wavenumber for different values of β_c and θ. Figures 8.12 through 8.14 correspond to a three-component plasma with electrons (e), core protons (c) and beam protons (b). For these figures, unless otherwise noted, $n_b = 0.10n_e$, $T_e = 1.50T_c$, $T_b = 0.10T_c$, $T_{\perp j} = T_{\parallel j}$ for all components, $v_o = 12v_c$, and $v_A/c = 2.33 \times 10^{-3}$. In this figure the upper two panels show growth of the ion/ion cyclotron instability, whereas the bottom two panels illustrate growth of the ion/ion subcyclotron instability. The former mode is also very weakly unstable in the lower right-hand panel.

ratio (specifically, $T_e/T_c = 1.5$) and a relatively cool proton beam (here $T_b/T_c = 0.10$).

The four panels of Figure 8.12 show the dispersion properties of this instability for these parameters and $0 \leq \beta_c \ll 1$. At $\beta_c = 0$, as illustrated in the upper left-hand panel of the figure, the lowest frequency branch of the ion/ion cyclotron instability, like the electron/ion cyclotron instability

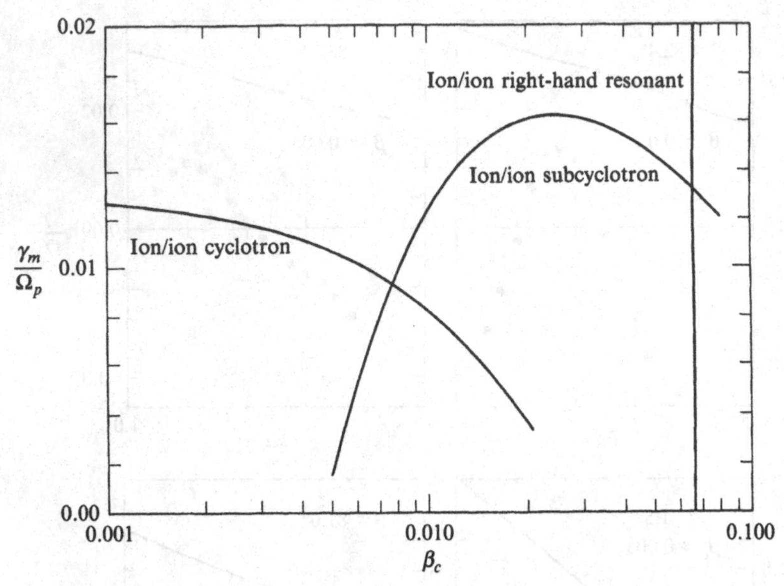

Fig. 8.13 The maximum growth rates of three ion/ion instabilities as functions of β_c.

(Figure 3.18), satisfies $1.0 < \omega_r/\Omega_p < 2.0$ and the growth rate is a small fraction of the proton cyclotron frequency. As β_c is increased from zero, the growth rate of the ion/ion cyclotron instability is reduced, and the propagation angle at maximum growth shifts to a more nearly perpendicular value of θ. At the same time, the introduction of nonzero β also introduces another regime of instability at less oblique propagation, longer wavelength and frequencies below the proton cyclotron frequency (Winske and Omidi, 1992). Further enhancements of β_c correspond to an increasing γ_m in this new regime, which we term the ion/ion subcyclotron instability; eventually this mode dominates its cyclotron counterpart, as suggested by the lower right-hand panel of Figure 8.12.

The maximum growth rates of these two ion/ion instabilities are illustrated as functions of β_c in Figure 8.13. This figure shows the decrease in γ_m of the cyclotron mode, the rise, domination and subsidence of the subcyclotron instability, and the eventual assertion of the ion/ion right-hand resonant instability when β_c becomes sufficiently large that the threshold of the electromagnetic instability (at $v_o/v_A = 2.18$ here) is exceeded. The maximum growth rate of the ion/ion subcyclotron instability has parameteric dependences that are similar to those of the ion/ion cyclotron instability: γ_m increases with n_b/n_e and decreases as T_b/T_c increases. So a number of

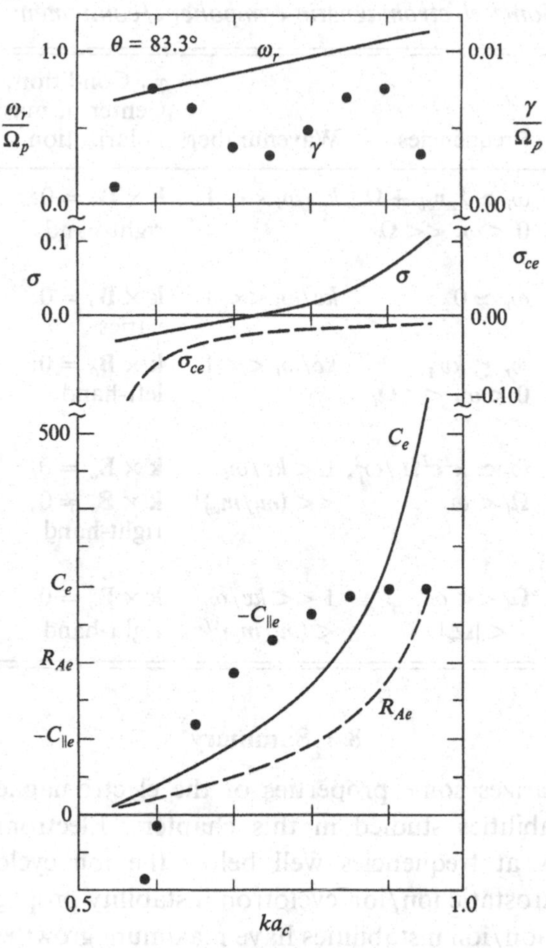

Fig. 8.14 The real frequency, growth rate and five transport ratios of the ion/ion subcyclotron instability ($ka_c \lesssim 0.75$) and the ion/ion cyclotron instability ($ka_c \gtrsim 0.75$) as functions of wavenumber. Here $\beta_c = 0.01$ and $\theta = 83.3°$.

the dispersion properties of the cyclotron and subcyclotron instabilities are similar.

In contrast, however, the transport properties of the two unstable modes are distinct, and may provide a basis for distinguishing between the two instabilities. Figure 8.14 shows the complex frequency and five dimensionless transport ratios for a choice of parameters at which the two modes co-exist. The helicities of the two modes bear different signs, and the electron compressibility, parallel compressibility, and Alfvén ratios of the subcyclotron instability are all significantly smaller in magnitude than the corresponding quantities of the cyclotron instability.

Table 8.1. *Some electromagnetic component/component instabilities*

Name	Frequencies	Wavenumbers	γ_m Condition; Center of mass polarization	Parameter regimes
Ion/ion right-hand resonant	$\omega_r \simeq k_z v_{ob} + \Omega_i$, $0 < \omega_r << \Omega_i$	$kc/\omega_i << 1$	$\mathbf{k} \times \mathbf{B}_o = 0$; right-hand	$v_A \lesssim v_o$ and very small n_b/n_e
Ion/ion nonresonant	$\omega_r \simeq 0$	$kc/\omega_i << 1$	$\mathbf{k} \times \mathbf{B}_o = 0$; varies	$v_A << v_o$ or modest n_b/n_e
Ion/ion left-hand resonant	$\omega_r \lesssim kv_A$, $0 < \omega_r << \Omega_i$	$kc/\omega_i << 1$	$\mathbf{k} \times \mathbf{B}_o = 0$; left-hand	Warm or hot beam and $v_A \lesssim v_o$
Electron/ion whistler	$\omega_r \simeq k^2c^2\Omega_i/\omega_i^2$, $\Omega_i < \omega_r$	$1 < kc/\omega_i << (m_i/m_e)^{1/2}$	$\mathbf{k} \times \mathbf{B}_o = 0$, $\mathbf{k} \times \mathbf{B}_o \neq 0$; right-hand	$v_A << v_o$
(Electron/electron) whistler heat flux	$\Omega_i << \omega_r << \lvert\Omega_e\rvert$	$1 << kc/\omega_i < (m_i/m_e)^{1/2}$	$\mathbf{k} \times \mathbf{B}_o = 0$; right-hand	$v_A << v_o$

8.6 Summary

Table 8.1 summarizes some properties of the electromagnetic component/component instabilities studied in this chapter. Electromagnetic ion/ion instabilities grow at frequencies well below the ion cyclotron frequency, whereas the electrostatic ion/ion cyclotron instability propagates at $\omega_r > \Omega_i$. Electromagnetic ion/ion instabilities have maximum growth rates at $\mathbf{k} \times \mathbf{B}_o = 0$, whereas the electrostatic ion/ion cyclotron instability propagates at angles strongly oblique to the magnetic field. Wavelengths of electromagnetic component/component instabilities scale as the ion or electron inertial length, whereas the wavelengths of electrostatic cyclotron instabilities scale as the ion or electron gyroradius. Relative drifts of electromagnetic component/component instabilities must typically exceed the Alfvén speed, in contrast to the thermal or ion acoustic speed threshold of electrostatic instabilities.

At $\beta \ll 1$, $c_s < v_A$, and electrostatic ion/ion and electron/ion instabilities with their lower thresholds are likely to be the dominant growing modes. As the plasma β increases, v_A becomes comparable to or less than c_s and electromagnetic component/component instabilities become the modes of lowest threshold. Thus electromagnetic component/component instabilities, like their electromagnetic temperature anisotropy counterparts, are favored by relatively large β values, and are most often found in high-β space environments such as the solar wind.

References

Abraham-Shrauner, B., and W. C. Feldman, Whistler heat flux instability in the solar wind with bi-Lorentzian velocity distribution functions, *J. Geophys. Res.*, **82**, 1889, 1977.

Akimoto, K., S. P. Gary and N. Omidi, Electron/ion whistler instabilities and magnetic noise bursts, *J. Geophys. Res.*, **92**, 11209, 1987.

Barnes, A., Theory of generation of bow-shock-associated hydromagnetic waves in the upstream interplanetary medium, *Cosmic Electrodynamics*, **1**, 90, 1970.

Beinroth, H. J., and F. M. Neubauer, Properties of whistler mode waves between 0.3 and 1.0 AU from Helios observations, *J. Geophys. Res.*, **86**, 7755, 1981.

Brinca, A. L., and B. T. Tsurutani, Survey of low-frequency electromagnetic waves stimulated by two coexisting newborn species, *J. Geophys. Res.*, **91**, 48, 1988.

Brinca, A. L., and B. T. Tsurutani, The oblique behavior of low-frequency electromagnetic waves excited by newborn cometary ions, *J. Geophys. Res.*, **94**, 3, 1989a.

Brinca, A. L., and B. T. Tsurutani, On the excitation of cyclotron harmonic waves by newborn heavy ions, *J. Geophys. Res.*, **94**, 5467, 1989b.

Coroniti, F. V., C. F. Kennel, F. L. Scarf and E. J. Smith, Whistler mode turbulence in the disturbed solar wind, *J. Geophys. Res.*, **87**, 6029, 1982.

Dum, C. T., E. Marsch and W. Pilipp, Determination of wave growth from measured distribution functions and transport theory, *J. Plasma Phys.*, **23**, 91, 1980.

Feldman, W. C., J. R. Asbridge, S. J. Bame, M. D. Montgomery and S. P. Gary, Solar wind electrons, *J. Geophys. Res.*, **80**, 4181, 1975.

Feldman, W. C., J. R. Asbridge, S. J. Bame, S. P. Gary and M. D. Montgomery, Electron parameter correlations in high-speed streams and heat flux instabilities, *J. Geophys. Res.*, **81**, 2377, 1976a.

Feldman, W. C., J. R. Asbridge, S. J. Bame, S. P. Gary, M. D. Montgomery and S. M. Zink, Evidence for the regulation of solar wind heat flux at 1 AU, *J. Geophys. Res.*, **81**, 5207, 1976b.

Fuselier, S. A., M. F. Thomsen, S. P. Gary, S. J. Bame, C. T. Russell and G. K. Parks, The phase relationship between gyrophase-bunched ions and MHD-like waves, *Geophys. Res. Lett.*, **13**, 60, 1986.

Galvez, M., S. A. Fuselier, S. P. Gary, M. F. Thomsen and D. Winske, Alpha particle heating in hot diamagnetic cavities, *J. Geophys. Res.*, **95**, 11975, 1990.

Gary, S. P., Ion-acoustic-like instabilities in the solar wind, *J. Geophys. Res.*, **83**, 2504, 1978.

Gary, S. P., Electromagnetic ion beam instabilities: hot beams at interplanetary shocks, *Astrophys. J.*, **288**, 342, 1985a.

Gary, S. P., Electromagnetic electron beam instabilities: hot, isotropic beams, *J. Geophys. Res.*, **90**, 10815, 1985b.

Gary, S. P., Electromagnetic ion/ion instabilities and their consequences in space plasmas: a review, *Space Sci. Revs.*, **56**, 373, 1991.

Gary, S. P., and C. D. Madland, Electromagnetic ion instabilities in a cometary environment, *J. Geophys. Res.*, **93**, 235, 1988.

Gary, S. P., and R. L. Tokar, The second-order theory of electromagnetic hot ion beam instabilities, *J. Geophys. Res.*, **90**, 65, 1985.
Gary, S. P., and D. Winske, Computer simulations of electromagnetic instabilities in the plasma sheet boundary layer, *J. Geophys. Res.*, **95**, 8085, 1990.
Gary, S. P., W. C. Feldman, D. W. Forslund and M. D. Montgomery, Electron heat flux instabilities in the solar wind, *Geophys. Res. Lett.*, **2**, 79, 1975a.
Gary, S. P., W. C. Feldman, D. W. Forslund and M. D. Montgomery, Heat flux instabilities in the solar wind, *J. Geophys. Res.*, **80**, 4197, 1975b.
Gary, S. P., R. A. Gerwin and D. W. Forslund, Electromagnetic current instabilities, *Phys. Fluids*, **19**, 579, 1976.
Gary, S. P., C. W. Smith, M. A. Lee, M. L. Goldstein and D. W. Forslund, Electromagnetic ion beam instabilities, *Phys. Fluids*, **27**, 1852, 1984.
Gary, S. P., K. Akimoto and D. Winske, Computer simulations of cometary-ion/ion instabilities and wave growth, *J. Geophys. Res.*, **94**, 3513, 1989.
Gary, S. P., R. H. Miller and D. Winske, Pitch-angle scattering of cometary ions: computer simulations, *Geophys. Res. Lett.*, **18**, 1067, 1991.
Glassmeier, K.-H., A. J. Coates, M. H. Acuña, M. L. Goldstein, A. D. Johnstone, F. M. Neubauer and H. Rème, Spectral characteristics of low-frequency plasma turbulence upstream of comet P/Halley, *J. Geophys. Res.*, **94**, 37, 1989.
Goldstein, M. L., and H. K. Wong, A theory for low-frequency waves observed at comet Giacobini-Zinner, *J. Geophys. Res.*, **92**, 4695, 1987.
Goldstein, M. L., C. W. Smith and W. H. Matthaeus, Large amplitude MHD waves upstream of the jovian bow shock, *J. Geophys. Res.*, **88**, 9989, 1983.
Goldstein, M. L., H. K. Wong, A. F. Vinas and C. W. Smith, Large-amplitude MHD waves upstream of the jovian bow shock: reinterpretation, *J. Geophys. Res.*, **90**, 302, 1985.
Gosling, J. T., M. F. Thomsen, S. J. Bame and C. T. Russell, On the source of diffuse, suprathermal ions observed in the vicinity of the Earth's bow shock, *J. Geophys. Res.*, **94**, 3555, 1989.
Kennel, C. F., F. L. Scarf, F. V. Coroniti, R. W. Fredricks, D. A. Gurnett and E. J. Smith, Correlated whistler and electron plasma oscillation bursts detected on ISEE-3, *Geophys. Res. Lett.*, **7**, 129, 1980.
Krauss-Varban, D., and N. Omidi, Structure of medium Mach number quasi-parallel shocks: upstream and downstream waves, *J. Geophys. Res.*, **96**, 17715, 1991.
Marsch, E., and T. Chang, Electromagnetic lower hybrid waves in the solar wind, *J. Geophys. Res.*, **88**, 6869, 1983.
Miller, R. H., T. I. Gombosi, S. P. Gary and D. Winske, The directional dependence of magnetic fluctuations generated by cometary ion pickup, *J. Geophys. Res.*, **96**, 9479, 1991.
Neugebauer, M., A. J. Coates and F. M. Neubauer, Comparison of picked-up protons and water-group ions upstream of comet Halley's bow shock, *J. Geophys. Res.*, **95**, 18745, 1990.
Schwartz, S. J., Plasma instabilities in the solar wind: a theoretical review, *Revs. Geophys.*, **18**, 313, 1980.
Sentman, D. D., J. P. Edmiston and L. A. Frank, Instabilities of low frequency, parallel propagating electromagnetic waves in the Earth's foreshock region, *J. Geophys. Res.*, **86**, 7487, 1981.
Smith, C. W., M. L. Goldstein, S. P. Gary and C. T. Russell, Beam driven ion

cyclotron harmonic resonances in the terrestrial foreshock, *J. Geophys. Res.*, **90**, 1429, 1985.

Thomas, V. A., and S. H. Brecht, Evolution of diamagnetic cavities in the solar wind, *J. Geophys. Res.*, **93**, 11341, 1988.

Thomsen, M. F., Upstream suprathermal ions, in *Collisionless Shocks in the Heliosphere: Reviews of Current Research*, Geophysical Monograph 35, American Geophysical Union, 1985.

Thomsen, M. F., J. T. Gosling, S. J. Bame, K. B. Quest, C. T. Russell and S. A. Fuselier, On the origin of hot diamagnetic cavities near the Earth's bow shock, *J. Geophys. Res.*, **93**, 11311, 1988.

Thorne, R. M., and B. T. Tsurutani, Resonant interactions between cometary ions and low frequency electromagnetic waves, *Planet. Space Sci.*, **35**, 1501, 1987.

Tokar, R. L., D. A. Gurnett and W. C. Feldman, Whistler mode turbulence generated by electron beams in Earth's bow shock, *J. Geophys. Res.*, **89**, 105, 1984.

Tsurutani, B. T., and E. J. Smith, Strong hydromagnetic turbulence associated with comet Giacobini-Zinner, *Geophys. Res. Lett.*, **13**, 259, 1986.

Tsurutani, B. T., D. E. Page, E. J. Smith, B. E. Goldstein, A. L. Brinca, R. M. Thorne and H. Matsumoto, Low-frequency plasma waves and ion pitch angle scattering at large distances($> 3.5 \times 10^5$ km) from Giacobini-Zinner: Interplanetary magnetic field α dependences, *J. Geophys. Res.*, **94**, 18, 1989.

Watanabe, Y., and T. Terasawa, On the excitation mechanism of the low-frequency upstream waves, *J. Geophys. Res.*, **89**, 6623, 1984.

Winske, D., and S. P. Gary, Electromagnetic instabilities driven by cool heavy ion beams, *J. Geophys. Res.*, **91**, 6825, 1986.

Winske, D. and M. M. Leroy, Diffuse ions produced by electromagnetic ion beam instabilities, *J. Geophys. Res.*, **89**, 2673, 1984.

Winske, D., and N. Omidi, Electromagnetic ion/ion cyclotron instability: theory and simulations, *J. Geophys. Res.*, **97**, 14779, 1992.

Wong, H. K., and M. L. Goldstein, Proton beam generation of oblique whistler waves, *J. Geophys. Res.*, **93**, 4110, 1988.

Wu, C. S., and R. C. Davidson, Electromagnetic instabilities produced by neutral-particle ionization in interplanetary space, *J. Geophys. Res.*, **77**, 5399, 1972.

Wu, C. S., X. T. He and C. P. Price, Excitation of whistlers and waves with mixed polarization by newborn cometary ions, *J. Geophys. Res.*, **93**, 3949, 1988.

Appendix A
The plasma dispersion function

Because many zeroth-order distribution functions are Maxwellian or Maxwellian-like, it is convenient to define the plasma dispersion function

$$Z(\zeta) \equiv \frac{1}{\sqrt{\pi}} \int_{-\infty}^{\infty} \frac{dx \exp(-x^2)}{x-\zeta} \tag{A.1}$$

where ζ is a complex variable and the integration follows the Landau contour; that is, it passes below the singularity at $x = \zeta$. Its analytic and numerical properties are discussed in Fried and Conte (1961). It is also related to the error function of complex argument which is treated in Abramowitz and Stegun (1964).

We here summarize some useful properties of this function. There are two alternative integral definitions which do not involve the Landau contour:

$$\int_0^{\infty} dt \exp(-\alpha^2 t^2 \pm i\beta t) = -\frac{i}{2|\alpha|} Z\left(\frac{\pm\beta}{2|\alpha|}\right), \tag{A.2}$$

$$Z(\zeta) = 2i \, \exp(-\zeta^2) \int_{-\infty}^{i\zeta} dt \exp(-t^2). \tag{A.3}$$

From the above definitions, it follows that

$$Z'(\zeta) = -2[1 + \zeta Z(\zeta)], \tag{A.4}$$

$$Z(\zeta^*) = -[Z(-\zeta)]^*, \tag{A.5}$$

$$Z(\zeta) + Z(-\zeta) = 2i\sqrt{\pi} \exp(-\zeta^2). \tag{A.6}$$

The small argument series expansion is

$$Z(\zeta) = i\sqrt{\pi} \exp(-\zeta^2) - 2\zeta[1 - 2\zeta^2/3 + 4\zeta^4/15 - ...] \quad (|\zeta| \ll 1). \tag{A.7}$$

The asymptotic expansion is

$$Z(\zeta) \simeq i\sqrt{\pi}\sigma \exp(-\zeta^2) - \frac{1}{\zeta}\left[1 + \frac{1}{2\zeta^2} + \frac{3}{4\zeta^4} + \ldots\right] \qquad (|\zeta| \gg 1) \qquad \text{(A.8)}$$

where

$$\sigma = 0 \quad \text{if} \quad \text{Im}\zeta > 0,$$

$$\sigma = 1 \quad \text{if} \quad \text{Im}\zeta = 0,$$

$$\sigma = 2 \quad \text{if} \quad \text{Im}\zeta < 0.$$

References

Abramowitz, M., and I. A. Stegun, eds., *Handbook of Mathematical Functions,* National Bureau of Standards, Washington, DC, 1964.

Fried, B. D., and S. P. Conte, *The Plasma Dispersion Function,* Academic Press, 1961.

Appendix B
Unperturbed orbits

Throughout this book, we use the properties of an unperturbed orbit, the position and velocity of a charged particle subject to the zeroth-order forces of a particular configuration. More specifically, to solve for the distribution function as a function of the independent variables $\mathbf{x}$, $\mathbf{v}$ and t, we often integrate over the time variable t' with dependent variables $\mathbf{x}' = \mathbf{x}(t')$ and $\mathbf{v}' = \mathbf{v}(t')$ determined from the solution of $\mathbf{F}_j = m_j\mathbf{a}_j(t')$ where $\mathbf{F}_j$ is that given as the coefficient of the $\cdot\partial f_j^{(1)}/\partial\mathbf{v}$ term in the linear kinetic equation.

Imposing the initial conditions $\mathbf{x}(t') = \mathbf{x}(t)$ and $\mathbf{v}(t) = \mathbf{v}(t')$, the unperturbed orbit for a charged particle in a uniform magnetic field $\mathbf{B}_o = \hat{\mathbf{z}}B_o$ is

$$v_x(t') = v_\perp \cos(\Omega_j\tau - \phi) \tag{B.1}$$

$$v_y(t') = -v_\perp \sin(\Omega_j\tau - \phi) \tag{B.2}$$

$$v_z(t') = v_z \tag{B.3}$$

and

$$x(t') = x(t) + (v_\perp/\Omega_j)[\sin(\Omega_j\tau - \phi) + \sin\phi] \tag{B.4}$$

$$y(t') = y(t) + (v_\perp/\Omega_j)[\cos(\Omega_j\tau - \phi) - \cos\phi] \tag{B.5}$$

$$z(t') = z(t) + v_z\tau \tag{B.6}$$

where $\tau \equiv t' - t$.

From these orbits, one may construct many different constants of that motion that are independent of t'. Three of the most useful are

$$v_z' \tag{B.7}$$

$$(v_x')^2 + (v_y')^2 \tag{B.8}$$

and

$$x' + v_y'/\Omega_j. \tag{B.9}$$

Zeroth-order distributions which are functions only of such constants of the motion are independent of t' and therefore satisfy the zeroth order Vlasov equation.

If a steady, uniform electric field $\mathbf{E} = \hat{\mathbf{z}}E_o$ is present, the perpendicular components of the orbit are unchanged, but

$$v_z(t') = v_z + \frac{e_j E_o \tau}{m_j}$$

and

$$z(t') = z(t) + v_z\tau + \frac{e_j E_o \tau^2}{2m_j}.$$

If a steady uniform electric field $\mathbf{E} = -\hat{\mathbf{x}}E_o$ is present, the x- and z-components of motion are unchanged from Equations (B.1) through (B.6), but

$$v_y(t') = -v_\perp \sin(\Omega_j\tau - \phi) + v_E \tag{B.10}$$

and

$$y(t') = y(t) + (v_\perp/\Omega_j)[\cos(\Omega_j\tau - \phi) - \cos\phi] + v_E\tau \tag{B.11}$$

where the $\mathbf{E}\times\mathbf{B}$ drift velocity is $\mathbf{v}_E = \hat{\mathbf{y}}v_E$ with $v_E = cE_o/B_o$. In this case, v_z' and $x' + v_y'/\Omega_j$ remain constants of the motion, but the constant associated with perpendicular kinetic energy becomes

$$(v_x')^2 + (v_y' - v_E)^2. \tag{B.12}$$

In the case of a uniform plasma acceleration $\mathbf{g} = -\hat{\mathbf{x}}g$, the gravitational drift of the jth species is $\mathbf{v}_{gj} = \hat{\mathbf{y}}v_{gj}$ with $v_{gj} \equiv g/\Omega_j$. Then Equations (B.10) through (B.12) apply if v_E is replaced by v_{gj}.

Appendix C
Integral evaluation

There are two procedures for evaluating integrals of the type

$$A_j(\mathbf{k},\omega) \equiv \frac{i}{n_j}\int d^3v f_j^{(M)}(v)\int_{-\infty}^{0} d\tau\ \exp[ib_j(\tau,\omega)] \tag{C.1}$$

where $f_j^{(M)}(v)$ is given by Equation (2.1.1) and $b_j(\tau,\omega)$ by Equation (2.3.8).

The first method utilizes cylindrical coordinates in velocity [Equation (2.3.8b)] and does the τ integration first. Use of the Bessel function identity

$$\exp(iz\cos\Phi) = \sum_{m=-\infty}^{\infty} i^m \exp(\pm im\Phi)J_m(z) \tag{C.2}$$

leads to

$$\int_{-\infty}^{0} d\tau\ \exp[ib_j(\tau,\omega)] = \sum_{m,n=-\infty}^{\infty} \frac{i^{m-n}\exp[i(m-n)\phi]J_m(k_y v_\perp/\Omega_j)J_n(k_y v_\perp/\Omega_j)}{i(k_z v_z - \omega - m\Omega_j)}. \tag{C.3}$$

Since $f_j^{(M)}(v)$ is independent of the azimuthal velocity angle ϕ, we next integrate with respect to this coordinate:

$$\int_0^{2\pi} d\phi \int_{-\infty}^{0} d\tau\ \exp[ib_j(\tau,\omega)] = \sum_{m=-\infty}^{\infty} \frac{2\pi J_m^2(k_y v_\perp/\Omega_j)}{i(k_z v_z - \omega - m\Omega_j)}. \tag{C.4}$$

The $v_\perp$ integration is accomplished through the use of the definite integral

$$\int_0^{\infty} dx\ x\ \exp(-\rho^2 x^2)J_p(\alpha x)J_p(\beta x) = \frac{1}{2\rho^2}\exp\left(-\frac{\alpha^2+\beta^2}{4\rho^2}\right) I_p\left(\frac{\alpha\beta}{2\rho^2}\right) \tag{C.5}$$

where $I_p(x)$ is the modified Bessel function of order p (Abramowitz and Stegun, 1964). If the v_z integration is taken to be over the Landau contour, $A_j(\mathbf{k},\omega)$ may be expressed in terms of the plasma dispersion function

(Appendix A):

$$A_j(\mathbf{k},\omega) = \frac{\exp(-\lambda_j)}{\sqrt{2}|k_z|v_j} \sum_{m=-\infty}^{\infty} I_m(\lambda_j) Z(\zeta_j^m) \tag{C.6}$$

where $\lambda_j \equiv (k_y a_j)^2$ and $\zeta_j^m \equiv (\omega + m\Omega_j)/\sqrt{2}|k_z|v_j$.

The alternate procedure expresses velocity in Cartesian coordinates, uses Equation (2.3.8a) and does the velocity integrations first. Assuming, as we do throughout, that velocity and time integrations are interchangeable,

$$A_j(\mathbf{k},\omega) = \frac{i}{n_j} \frac{1}{(2\pi v_j^2)^{3/2}} \int_{-\infty}^{0} d\tau \int dv_x dv_y dv_z \exp\left[-\frac{(v_x^2+v_y^2+v_z^2)}{2v_j^2}\right]$$

$$\exp\left[i\frac{k_y v_x}{\Omega_j}(\cos\Omega_j\tau - 1) + \frac{ik_y v_y}{\Omega_j}\sin\Omega_j\tau + k_z v_z \tau\right] \tag{C.7}$$

a threefold application of the definite integral

$$\int_{-\infty}^{\infty} dx\ \exp(-\alpha x^2 \pm i\beta x) = \left(\frac{\pi}{\alpha}\right)^{1/2} \exp\left(-\frac{\beta^2}{4\alpha}\right) \tag{C.8}$$

yields the Gordeyev integral form

$$A_j(\mathbf{k},\omega) = i\int_0^{\infty} d\tau\ V_j(\tau) \tag{C.9}$$

where

$$V_j(\tau) \equiv \exp(i\omega\tau)\exp[-\lambda_j(1-\cos\Omega_j\tau)]\exp(-k_z^2 v_j^2 \tau^2/2). \tag{C.10}$$

This form is particularly useful for deriving the transition to the unmagnetized form of $A_j(\mathbf{k},\omega)$. If $\gamma/\Omega_j \gg 1$ or $(k_z v_j/\Omega_j)^2 \gg 1$ so that one may expand $1 - \cos\Omega_j\tau \simeq (\Omega_j\tau)^2/2$ in (C.10), one obtains, by means of Equation (A.2),

$$A_j(\mathbf{k},\omega) = \frac{1}{\sqrt{2}kv_j} Z\left(\frac{\omega}{\sqrt{2}kv_j}\right). \tag{C.11}$$

In the more general case, one uses the identity

$$\exp(z\cos\theta) = \sum_{m=-\infty}^{\infty} \exp(im\theta) I_m(z) \tag{C.12}$$

in conjunction with Equation (A.2) to reduce (C.9) to (C.6).

Using a similar procedure, it follows that

$$\frac{i}{n_j}\int d^3v f_j^{(M)}(v)\mathbf{v} \int_{-\infty}^{0} d\tau\ \exp[ib_j(\tau,\omega)]$$

$$= \frac{\exp(-\lambda_j)}{\sqrt{2}|k_z|}\left\{-i\hat{\mathbf{x}}\frac{k_y v_j}{\Omega_j}\sum_{m=-\infty}^{\infty}[I_m(\lambda_j) - I'_m(\lambda_j)]Z(\zeta_j^m)\right.$$

$$\left.-\hat{\mathbf{y}}\frac{\Omega_j}{k_y v_j}\sum_{m=-\infty}^{\infty} mI_m(\lambda_j)Z(\zeta_j^m) - \frac{\hat{\mathbf{z}}}{\sqrt{2}}\frac{|k_z|}{k_z}\sum_{m=-\infty}^{\infty} I_m(\lambda_j)Z'(\zeta_j^m)\right\} \qquad \text{(C.13)}$$

and

$$\frac{i}{n_j}\int d^3v f_j^{(M)}(v)\int_{-\infty}^{0} d\tau\, \mathbf{v}' \exp[ib_j(\tau,\omega)]$$

$$= \frac{\exp(-\lambda_j)}{\sqrt{2}|k_z|}\left\{i\hat{\mathbf{x}}\frac{k_y v_j}{\Omega_j}\sum_{m=-\infty}^{\infty}[I_m(\lambda_j) - I'_m(\lambda_j)]Z(\zeta_j^m)\right.$$

$$\left.-\hat{\mathbf{y}}\frac{\Omega_j}{k_y v_j}\sum_{m=-\infty}^{\infty} mI_m(\lambda_j)Z(\zeta_j^m) - \frac{\hat{\mathbf{z}}}{\sqrt{2}}\frac{|k_z|}{k_z}\sum_{m=-\infty}^{\infty} I_m(\lambda_j)Z'(\zeta_j^m)\right\} \qquad \text{(C.14)}$$

References

Abramowitz, M., and I. A. Stegun, eds., *Handbook of Mathematical Functions*, National Bureau of Standards, Washington, DC, 1964, p. 374.

Index of symbols

Symbol	Description	Page where introduced
$\mathbf{A}$	magnetic vector potential	92
a_j	gyroradius of the jth species or component	5
$\mathbf{B}$	magnetic field	4
$\mathbf{B}_o$	background magnetic field	5
b	as subscript, denotes beam component	32
$b_j(\tau, \omega)$	$\mathbf{k} \cdot (\mathbf{x}' - \mathbf{x}) - \omega(t' - t)$	23
C_j	compressibility of the jth species or component	93
$C_{\parallel j}$	parallel compressibility of the jth species or component	94
c	speed of light	2
c	as subscript, denotes cool or core component	32
c_s	ion sound speed	18
$c_{s\perp}$	ion sound speed at propagation perpendicular to $\mathbf{B}_o$	103
$\mathsf{D}(\mathbf{k}, \omega)$	dispersion tensor for electromagnetic fluctuations	86
$\mathbf{E}$	electric field	4
$\mathbf{E}_o$	background electric field	6
$\mathbf{E_L}$	longitudinal fluctuating electric field	8
$\mathbf{E_T}$	transverse fluctuating electric field	8
e	as subscript, denotes electron species	3
e_j	charge of a jth species particle	2
$\hat{\mathbf{e}}_\pm$	$(\hat{\mathbf{x}} \pm i\hat{\mathbf{y}})/\sqrt{2}$	89
f	frequency	76
$f_j(\mathbf{x}, \mathbf{v}, t)$	distribution function of the jth species or component	2
$f_j^{(M)}$	Maxwellian distribution function of the jth species or component	11
$\mathbf{g}$	plasma acceleration	173
h	as subscript, denotes hot component	33
$I_p(\lambda_j)$	modified Bessel function of argument λ_j and order p	174
i	as subscript, denotes ion species	3
$\mathbf{J}$	current density	4
$J_p(z)$	Bessel function of argument z and order p	174
$K_j(\mathbf{k}, \omega)$	susceptibility of the jth species or component	13
$\mathbf{k}$	wavevector	1
k_B	Boltzmann factor	6
k_j	Debye wavenumber of the jth species or component	5
$\mathbf{k}_m$	wavevector of maximum growth rate	7
(M)	As superscript, denotes response functions calculated from Maxwellian distributions	11
m_j	mass of a jth species particle	2
n_j	particle density of the jth species or component	3
P	polarization of electromagnetic fluctuations	91
$\mathbf{P}_j$	momentum density of the jth species or component	3
p	as subscript, denotes proton species	3
$\mathbf{Q}_j(\mathbf{k}, \omega)$	response function vector defined by Equation (8.1.4)	144
$\mathbf{q}_j$	heat flux density of the jth species or component	4
$\mathbf{R}_j(\mathbf{k}, \omega)$	response function vector defined by Equation (8.1.5)	144

Index

Printed in the United States
By Bookmasters